This book is due for return on or before the last date shown below.

Don Gresswell Ltd., London, N.21   Cat. No. 1208

DG 02242/71

# Contents

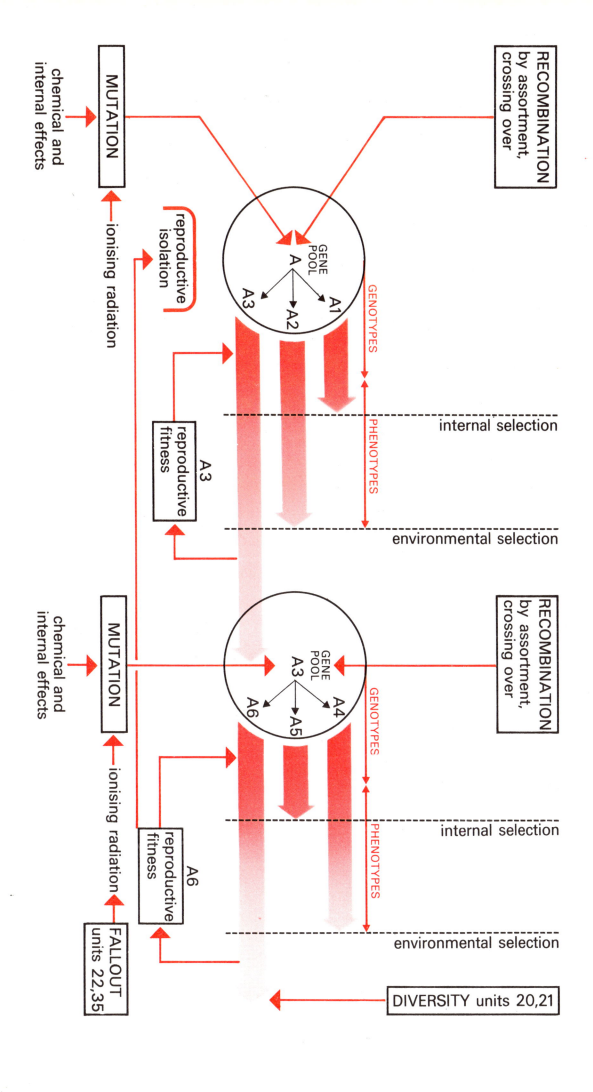

**Table A**

## List of Scientific Terms, Concepts and Principles used in Unit 19

| Taken as pre-requisites | | | Introduced in this Unit | | | |
|---|---|---|---|---|---|---|
| **1** | **2** | | **3** | | **4** | |
| Assumed from general knowledge | Introduced in a previous Unit | Unit No. | Developed in this Unit or in its set book(s) | Page No. | Developed in a later Unit | Unit No. |
| | DNA | 13 | adaptation | 9 | | |
| | genetic information | 17 | clone | 11 | | |
| | genetic continuity | 17 | dominant | 13 | | |
| | mitosis | 17 | recessive | 13 | | |
| | chromosomes | 17 | chromatid | 17 | | |
| | co-linearity | 17 | diploid | 17 | | |
| | enzyme | 10 | gonads | 17 | | |
| | polypeptide chain | 14 | haploid | 17 | | |
| | nucleotide | 14 | homologous chromosomes | 17 | | |
| | gametes | 17 | meiosis | 17 | | |
| | phenotype | 17 | crossing over | 22 | | |
| | genotype | 17 | assortment | 23 | | |
| | phage | 17 | polyploid | 23 | | |
| | | | recombination | 23 | | |
| | | | mutation | 25 | | |
| | | | population | 28 | | |
| | | | evolution | 29 | | |
| | | | inheritance of acquired characters | 30 | | |
| | | | natural selection | 31 | | |
| | | | micro-evolution | 36 | | |
| | | | selection | 36 | | |
| | | | heterozygous | 42 | | |
| | | | homozygous | 42 | | |
| | | | speciation | 47 | | |
| | | | species | 47 | | |
| | | | gene pool | 49 | | |
| | | | gene flow | 50 | | |
| | | | hybrid | 51 | | |
| | | | intelligence quotient | 56 | | |

*Scientific terms used in this Unit but not listed above are marked thus † and defined in the glossary (p. 61).*

## Objectives

When you have completed this Unit, you should be able to:

1 Define, or recognize the best definitions of, the terms listed in Table A, p. 5 (*SAQs* 1, 2, 3)

2 List, or select correct examples of:

(a) at least three agents which may cause a rise in the mutation rate in a population of living organisms;

(b) at least two examples of the frequency with which particular mutations are known to occur;

(c) two examples of the chemical change which has occurred in a phenotype as a result of a particular mutation of the genotype;

(d) three examples of the sort of changes in the structure of the DNA or chromosome which might constitute a mutation;

(e) four examples of adaptation in a population arising from selection;

(f) one example of the effect of conflicting selection pressures on the frequency of a gene in a population;

(g) at least one item of evidence that selection is acting on the human gene pool at the present time. (*Text Question (TQ)*, *SAQs* 4, 7)

3 (a) Draw or select annotated diagrams to show the essential differences in the behaviour and distribution of the chromosomes in meiotic and mitotic cell division.

(b) Interpret experimental data on the redistribution of characteristics following sexual reproduction, in terms of the behaviour of the chromosomes. (*TQ*)

4 Discuss in 500 words the potential evolutionary advantages to a species of sexual reproduction.

5 Given appropriate data, design simple experiments to:

(a) determine whether a variant in a population is phenotypic or genotypic;

(b) show whether a particular mutation occurs spontaneously or as a consequence of some specific action on the genotype by the environment;

(c) determine whether two characteristics are controlled by genes on the same or different chromosomes. (*TQ*)

6 To select and interpret data which give support for the hypothesis that under some circumstances natural selection is acting to favour stability and under others to favour change in a population. (*CMA*)

7 (a) To recognize the most important factors involved in distinguishing species. (*SAQ* 8)

(b) To recognize the circumstances under which new races may be formed, and from them, new species. (*Unit* 21, *section* 21.1)

8 By means of an essay of 1 000 words, or a suitable essay plan:

(a) outline Darwin's theory of evolution and compare its ability to explain the living world with that of earlier theories;

(b) construct an argument to support the idea that the evolution of species, and their adaptation to their environment, can be explained solely on the basis of selection acting on random variation;

6

(c) make a reasoned case for or against the statement 'that change in the human genotype is likely to be an important factor in future human evolution'. (*SAQ* 9)

9  Apply principles expanded in this and earlier Units to totally new (given) situations to which they are relevant. (*CMA*)

## General Aims

1  To show how genetic variations may arise within a species or a population, and how this variation, coupled with the over production of offspring, can lead to a change in the prevalent genotype of that species or population.

2  To consider species and speciation in terms of the theory of natural selection.

3  To induce the reader to consider a number of practical problems in the light of current views on the action of selection on populations.

# Study Comment

1 It is important that before you read this Unit, you should consider the objectives. As with the other Units in the course, it is only by looking at the objectives and the *SAQs* that you can know what we hope you will gain from it. They should give you a clear idea of what facts, ideas and principles we expect you to retain after completing it.

2 Brief summaries have been provided at the end of the section on mutation (19.2.1), the section on meiosis (19.2.3) and at the end of the Unit. You are asked to make your own very brief summaries of what you consider to be the important points, at the end of the sections on evolutionary theories (19.5), micro-evolution and adaptation (19.6) and speciation and gene pools (19.7). We have indicated what we consider these to be, and we hope there will be a reasonable amount of agreement between your version, our version and the objectives!

3 Quite early in section 19.2.3, when the basic mechanics of meiotic cell division have been introduced, you are asked to do the first parts of the Home Experiment, which involves the use of photographs rather than actual live material. We hope that this will satisfy you that the main points we make about meiosis are in fact true.

The set reading for this Unit is a paper by Dr. N. G. Smith on the reproductive isolation of various species of Arctic gulls, and you should read it when you have completed the sections on speciation and the isolation of gene pools.

4 If you care to pursue the topics raised in this Unit further, we recommend that you read T. Dobzhansky's *Evolution, Genetics and Man* (Chapman and Hall, Wiley, 1955). On a rather narrower front, *Mankind Evolving*, by the same author (Bantam Books, 1970), is a most interesting book, well worth your attention.

**.1 Introduction**

If you think back over the last two Units, 17 and 18, you will be aware of a paradox. Unit 18, particularly the television component of it, gave you some indication of the tremendous diversity of living organisms. There are indeed well over a million and a half distinct types, their bodily form ranging from the bacterial to the elephantine. Yet the message emerging from Unit 17 was one of continuity and sameness. The structure and function of a cell are determined largely by its proteins, and these depend on genetic factors, in fact on the precise sequence of bases in the DNA molecule. Mitosis, the process by which cells multiply, preserves this sequence. So the continuity of form and function conferred on the cell or organism would seem to be potentially everlasting, yet all around us we see diversity.

How has this arisen?

If we say that in the process of cell reproduction 'like begets like', it might seem reasonable to assume that all living organisms must have been present from the start of life. This indeed is the view held by Aristotle and Linnaeus among others. It requires that all organisms arose by a process of 'Special Creation', being produced, in unalterable form, by a Creator. Early versions of this theory assumed that they were all created together, at a time known as the Creation. After the discovery of the fossilized remains of hitherto unknown species in rocks known to be of different ages, it was suggested that creation was a sequential process, with intervals of catastrophe wiping out all life, followed by re-creation—a doctrine known as catastrophism.

'Special Creation'

catastrophism

An alternative view is that life may have arisen as just one or a few types of simple organism and that all the present and extinct forms have evolved from these. In this event it will be likely that such evolution will still be continuing. This general theory has, for nearly 100 years, had the support of almost all serious scientists. The ideas of 'Special Creation' are of course impossible to disprove; they are not testable scientific theories, but articles of faith. In this Unit, we are not setting out to prove that biological evolution has occurred, nor to examine its philosophical implications. Its occurrence is generally recognized, so here we will consider only the mechanism by which it has occurred.

However, there is something perhaps even more striking than the great diversity which living organisms show, and that is the extent to which they are precisely adapted to fit their environments. Any detailed study of an organism reveals the truly amazing degree to which its form, physiology, biochemistry and behaviour are exactly suited to the way it lives.

adaptation

For example, a plant may achieve sexual reproduction by producing a flower which is shaped and coloured in such a way as to lure just one particular species of insect. This insect must be flying at the time of year that the plant has ripe pollen, and the behaviour of the insect and the structure of the flower must be precisely co-ordinated if the method is to work. If the adaptation is not correct, it will be useless.

Fish living in cold water will require enzyme systems able to maintain all aspects of life at a low temperature, whereas the enzymes of a fish living in a tropical swamp must achieve the same results in water nearly as warm as a mammal's bloodstream. The tropical fish will perish from cold in a trout stream, as its enzymes will not be able to maintain essential functions

at a sufficiently fast rate. Equally, a trout will die in a tropical swamp, but in this case because there will be insufficient oxygen dissolved in the warm, stagnant water (its respiratory system is adapted to work in cold, but well-oxygenated water). The complex nature of these adaptations seems at first sight little short of miraculous, and it is hardly surprising that adaptation has often been taken as evidence of divine planning. Any scientific theory of evolution must therefore account satisfactorily for this fundamental phenomenon, exhibited by all living things.

No form of evolution can occur unless there is genetic change or variation, as well as the continuity discussed in Unit 17; thus the first section of this Unit deals with variation. We then consider the mechanism by which variation is exploited, and how this may lead to new species; finally, very briefly, we consider in what directions evolution appears to be going at the moment.

## .2  Variation

### .1  Variation caused by mutation

So far you have considered only one reproductive process, *mitosis* (Unit 17). As a prelude to mitosis each of the parental chromosomes duplicates to form two identical copies, or *chromatids*, one going to each daughter cell. So the daughter cells should be genetically identical, to each other and to the cell from which they arose. Many unicellular organisms, such as *Euglena* (TV programme of Unit 18), reproduce entirely by mitosis, or, as in the case of bacteria, by a similar form of 'simple division' not usually called mitosis because of the different arrangement of the genetic material within the cells. A succession of organisms produced mitotically from a single ancestor is sometimes referred to as a *clone*. The term can be applied to cells as well as to organisms, for example the cartilage cells mentioned in Unit 17. One would expect all members of a clone to be genetically identical, and so to be *potentially* identical in all respects—appearance, growth, metabolism etc.

In fact, the adults of genetically identical individuals may not be identical in all respects because they may have received rather different treatment by the environment. For example, human 'identical' twins are rarely quite indistinguishable, although they share the same patterns of DNA. One of them may have been ill when a baby and be smaller, or he may have a scar on his nose. His behaviour may be rather different: he may be afraid of horses, having had his nose bitten by one.

However, unicellular organisms can be grown in a laboratory under almost standard conditions, so that they receive substantially the same treatment from the environment. For example, bacteria may be cultured on a specially prepared nutrient jelly in covered glass dishes in an incubator. Careful mixing of the jelly ensures that they all receive the same foodstuffs in the same concentrations. The incubator ensures that the temperature, light and humidity are constant for all the individuals at all stages of their growth. If the cultures are set up under sterile conditions, the bacteria will only have each other for company; there will be no other organisms present which might affect some individuals and not others. Under these conditions you might expect that all the individuals of a clone would be indistinguishable from one another. There are many ways in which this may be tested. One such way is to take *Escherichia coli* (*E. coli*), a common bacterium from the gut of man and many other mammals, and grow a clone at 35° C in the standard culture dish known as a Petri dish. Under optimum conditions these organisms will divide every 20–30 minutes and they will be well distributed across the dish by the time they number a few billion (say $3 \times 10^{10}$). When examined microscopically they will appear identical.

mitotis

clone

genetically identical individuals

11

Another way to examine them, however, would be to add a suspension of T4 bacteriophage to the cultures.

**From what you know from Unit 17, what would you expect to happen to the culture?**

The phage will attack the bacteria and destroy the culture.

When examined by eye, the growing colony will appear as a large stain, which spreads across the jelly from the point of inoculation. When the phage is added, the colony will apparently be wiped out. But within a few hours one or a small number of spots will appear on the plate and begin to spread. These represent the start of new colonies, arising from one or more bacteria which were not killed by the phage. The resistance of the survivors is clearly heritable, for by the time they can be seen with the naked eye the new colonies number thousands of individuals, all growing in a medium containing large numbers of T4 phage.

heritable resistance

Obviously you would have been mistaken to think that all the individuals of the original clone were identical; a very few of them, perhaps one in every hundred million, must have possessed a striking difference—immunity to attack by T4 phage. Similar experiments using the antibiotic streptomycin instead of phage, again show that the bacteria are not all identical; in this case a few individuals are resistant to the antibiotic. These experiments raise a number of interesting points, and will be mentioned again on pp. 33–4.

All the bacteria are descended from the same ancestor by mitotic division; yet they are not all identical. What has happened? Somewhere in the sequence of divisions a sudden, sharp change has occurred: a genotype which does not confer immunity against T4 phage or streptomycin has changed to one which does. The change may be said to be sudden in this case because there is no evidence for any state of partial or transient resistance which precedes full resistance. A change of this type to the genotype is called a *mutation*. Precisely what a mutation involves chemically is now becoming clear in a few instances. An enormous amount of research into the genetics of *E. coli* has been performed in the last twelve years; indeed a fair amount of the total knowledge of chemical genetics is based on the genetics of this bacterium and of the phages which attack it. (In the discussion on co-linearity in Unit 17, the examples used were in fact mutants of the T4 phage.)

mutation

From what was said in Unit 17, it is clear that a change in the genotype will almost certainly mean a change in the configuration of the nucleic acid. Obviously, a large change in the template is likely to have a drastic effect on the phenotype; it is, however, important to try and judge what is the basic or minimum unit of change that could be called a mutation. If we accept the definition of a gene as being that genetic unit which controls the production of one polypeptide chain, then clearly a mutation may involve only one gene, or even a small part of one gene.

gene

This is of course implied by the theory of the 'triplet code', but direct evidence has been hard to come by. If the theory is correct—and the evidence for co-linearity goes far to support it—a change in one nucleotide may result in the incorporation of a different amino acid into the protein finally produced, and this in turn may profoundly affect the nature and function of that protein, and thus the cell.

triplet code

For example, some mutants of *E. coli* cannot grow unless the medium they are in contains the amino acid tryptophan. The original, or 'wild-type', can manufacture its own tryptophan, provided that the amino acid serine is present. Mutants that cannot do this turn out to have a defect in the protein of the enzyme that is used for the conversion. In one case,

12

the defect is caused by the replacement of one molecule of the amino acid glycine by a molecule of arginine at a particular point in the protein. Thus a functional change in an organism has been shown to be due to the change of a single amino acid in a protein made up of 267 amino acids.

The speed with which *E. coli* can be bred made it possible for workers in Europe and America to identify several other changes in the same enzyme, and relate them to specific changes in the bacterial DNA (Fig. 1). It emerged from this work that not only do the positions of the mutations on the DNA correspond with the sites of the changed amino acids on the polypeptide chain (providing further evidence of co-linearity) but that by comparing these amino acids with their known codons (Unit 17) it can be deduced that each mutation *could* indeed result from a change in a single nucleotide. This would be in line with predictions made from the 'triplet code' theory. It is a remarkable thought that a visible, functional change in a whole organism may result from the changing of a single nucleotide on a strand of DNA.

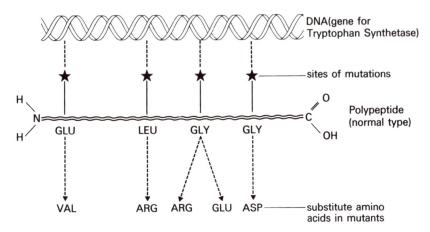

*Figure 1 Showing co-linearity of the sites of changes on the bacterial DNA with those of substituted amino acids in the protein of the enzyme tryptophan synthetase.*

But, although this work gave the first evidence that a mutation might involve as little as a single nucleotide on a strand of DNA, the first demonstration that a visible change in the phenotype could be traced to just one amino-acid change in a protein was achieved earlier than this, by Ingram in 1957. In this case, the discovery did not lend itself so well to fundamental investigation, but in many ways was even more interesting. Since about 1950, a good deal of interest had been taken in a disease known as sickle-cell anaemia. This is an often fatal 'anaemia' (anaemia in the sense that it is a failure of the red blood cells) which is relatively common in Africans, particularly West Africans. Whenever the blood cells of a victim encounter a low level of oxygen, as in the tissues or venous system, they are liable to collapse into a sickle shape and may form blockages and other complications in the blood vessels.

sickle-cell anaemia

It was found that this disease is inherited, as a single mutant *recessive gene**. The consequence of a child inheriting the sickling gene from both parents is that it has only a 20 per cent chance of surviving to maturity.

---

* The term recessive in a genetic context means that the character determined by the gene —in this case the disease—only shows itself in the phenotype if it appears in the relevant chromosomes from both parents. If it appears in just one, with the normal gene opposite, it does not affect the phenotype. The character determined by the normal gene is then said to be dominant.

Where an identical gene has been inherited from both parents the individual is said to be homozygous for this gene. Where the genes on the corresponding points of the relevant chromosomes are not identical, for example where a mutant is opposite to the normal gene, the individual is said to be heterozygous for this gene. Thus a sufferer from sickle-cell anaemia must be homozygous for the sickling gene, as it is recessive.

It was discovered in 1949 that the difference between sickling and normal red blood cells was in the haemoglobin they contained. Haemoglobin (Hb) molecules are constructed of two different types of polypeptide chains, α and β. Each chain contains about 150 amino acids. Two of each type of chain are present in each haemoglobin molecule.

The manufacture of the α chain is controlled by a different gene from that of the β chain, so that a single mutation will affect either one pair of chains or the other, not both. The haemoglobin in the sickling cells is known as Haemoglobin S (Hb.S). Between 1956–59, Ingram was able to show that the α chains of Hb.S were perfectly normal, as indeed were the β chains except for one amino acid. In position 6 on the chain, normal Hb has glutamic acid, but in Hb.S it has been replaced by valine. Thus, the only difference in the primary structure of a molecule with a total of nearly 600 amino acids (4 chains of 150) is that two glutamic acid molecules have been replaced by two molecules of valine (Fig. 2). Yet this difference has killed tens of thousands of people. It seems that the small difference produced by the substitution of two valines for the two glutamic acid molecules normally present, results in profound changes in the folding of the molecule, thus changing its shape. This in turn affects the behaviour of the molecule, particularly its solubility, and seems to result in the Hb molecules clumping together to form long helices when they lose oxygen to the tissues. Thus, apart from the dangers mentioned above, the haemoglobin becomes very inefficient as an oxygen carrier.

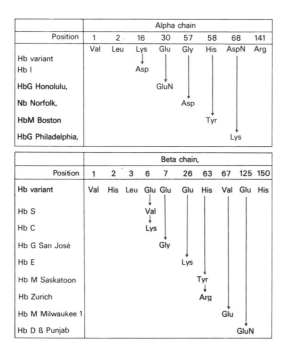

| | Alpha chain | | | | | | | |
|---|---|---|---|---|---|---|---|---|
| Position | 1 | 2 | 16 | 30 | 57 | 58 | 68 | 141 |
| Hb variant | Val | Leu | Lys | Glu | Gly | His | AspN | Arg |
| Hb I | | | Asp | | | | | |
| HbG Honolulu, | | | | GluN | | | | |
| Nb Norfolk, | | | | | Asp | | | |
| HbM Boston | | | | | | Tyr | | |
| HbG Philadelphia, | | | | | | | Lys | |

| | Beta chain, | | | | | | | | | |
|---|---|---|---|---|---|---|---|---|---|---|
| Position | 1 | 2 | 3 | 6 | 7 | 26 | 63 | 67 | 125 | 150 |
| Hb variant | Val | His | Leu | Glu | Glu | Glu | His | Val | Glu | His |
| Hb S | | | | Val | | | | | | |
| Hb C | | | | Lys | | | | | | |
| Hb G San José | | | | | Gly | | | | | |
| Hb E | | | | | | Lys | | | | |
| Hb M Saskatoon | | | | | | | Tyr | | | |
| Hb Zurich | | | | | | | Arg | | | |
| Hb M Milwaukee 1 | | | | | | | | Glu | | |
| Hb D ß Punjab | | | | | | | | | GluN | |

*Figure 2   Positions of substituted amino acids in various mutant forms of haemoglobin.*

The existence of this disease has other implications which will be explained in section 19.6.4, but at this point Hb.S is interesting as the first example of a single gene mutation being shown to affect just one amino acid of a polypeptide chain. Since then other mutated forms of Hb have also been shown to differ by just one amino acid (Fig. 2).

## Causes of mutation

It seems that anything which changes the sequence of nucleotides along the DNA molecule could cause a mutation. This change may be merely the substitution of one nucleotide for another, as in the cases considered above, or it may involve several nucleotides. It may also be caused by the deletion of one or more from the sequence altogether, or the addition of completely new ones. Some mutations can certainly be shown to be due to additions or deletions, but it seems that this generally results in the production of a completely non-functional protein.

Mutations can be produced experimentally in the laboratory, by using various different agents. For example, a large number of *chemicals* can be shown to induce mutation. (The actions of some of them are discussed in the black-page Appendix to this section.) They may act to cause deletions, or simply substitutions of single bases.

**mutation and chemicals**

Rather surprisingly, the *temperature* at which some animals, such as the fruit-fly *Drosophila* (film strip 19/20a), are living may have an effect: as the temperature of their environment rises, so does the rate at which mutations occur. This is probably not a relevant factor in animals that maintain a nearly constant body temperature.

**mutation and temperature**

Another factor that may induce mutation is the absorption of *ionizing radiation* such as X-rays or accelerated particles from a radioactive source. Irradiation of tissues may result in the breaking of some of the bonds within the DNA. When these are made good, substitutions may occur. This connection between radiation and mutation has led to controversy over the release into the environment of radioactive matter by military and commercial undertakings. You can find more detail of all three of these causes of mutation in Appendix 1 (Black).

**mutation and ionizing radiation**

## Mutation rates

There is, as yet, insufficient evidence to say precisely what causes most naturally occurring mutations. It seems likely, however, that the main cause is chemical. Background radiation is too slight under normal conditions to account for all of the observed mutation rate (at least that observed in *Drosophila*). Also, as it has a cumulative effect, a human with a generation time of nearly 30 years would be expected to accumulate some 360 times as many radiation-induced mutations as *Drosophila*, with a generation time of only four weeks. However, the overall mutation rate in humans seems to be only a little higher than that of *Drosophila*.

In spite of the general state of ignorance about the natural causes of mutation, there is quite a lot of information about the mutation rates of particular genes.

The rate of mutation to T4 phage resistance in *E. coli* is of the order of one in every hundred million, i.e. the chance of it occurring in any one bacterium is $10^{-8}$. It has been estimated that in the T4 phage itself the average rate of detectable mutation is $10^{-6}$ per gene. However there is no reason to suppose that all genes within an organism tend to mutate with the same frequency. Indeed it can be shown in several organisms, including the T4 phage, that there are 'hot spots', i.e. some genes or groups of nucleotides within genes that mutate far more frequently than others.

In maize plants, which live for just one season, seven genes have been compared for frequency of mutation. The rates varied from 492 per million for the colour factor 'R', to less than one per million for the factor 'Wx' (waxy seeds).

**Table 1  Mutations observed in seven genes of maize**

| Gene | Individuals examined | Mutations observed | Mutations per 1 000 000 individuals |
|---|---|---|---|
| R (colour factor) | 554 786 | 273 | 492 |
| I (colour inhibition) | 265 391 | 28 | 106 |
| $P_2$ (purple colour) | 647 102 | 7 | 11 |
| Su (sugar) | 1 678 736 | 4 | 2.4 |
| Y (yellow seeds) | 1 745 280 | 4 | 2.2 |
| Sh (shrunken seeds) | 2 469 285 | 3 | 1.2 |
| Wx (waxy seeds) | 1 503 744 | 0 | 0 |

In humans, mutations to the gene causing haemophilia, the disease in which blood clotting is impaired, arise 20 to 30 times in every million individuals. The mutation responsible for producing achondroplastic dwarfism† apparently occurs more often—50–100 times per million. Whilst this is within the range mentioned above for maize, it is much higher than that for the bacteria and phages, with 30-minute generation times. However, comparisons between such different organisms are rather hard to interpret in the present state of knowledge.

### 19.2.2  Summary of section 19.2.1

1  The occurrence of mutations can be demonstrated in a number of ways, one of which is to grow a clone of normal *E. coli*, and subject it to massive attack by T4 bacteriophage. If resistant individuals have arisen by mutation, these will found new colonies, which can be counted.

2  Several other mutants of *E. coli* cannot synthesize tryptophan from serine. It is known that the only deficiency in the bacterium is the replacement of one amino acid in an enzyme by another one, but this is enough to make an important change in the bacterium. The points on the bacterial DNA where the mutations have occurred have been identified, and in at least one case the mutation may involve no more than one nucleotide of the DNA.

3  The disease 'sickle-cell anaemia' is known to be caused by a single, recessive gene. The only difference between normal haemoglobin and the mutant form lies in the replacement of one amino acid by another in two of the four polypeptide chains making up the molecule.

4  Three of the factors which may cause mutation are listed, and the rate at which some particular genes are known to mutate are given.

5  From the above, we can say that mutations occur, that the frequency with which some of them occur is known, that some of the agents which induce them are known, and that in at least one case there is evidence that only one nucleotide is involved. In at least two cases, the detectable changes produced in the body of an organism can be attributed to the replacement of just one amino acid of a polypeptide chain.

### 19.2.3  Variation caused by meiosis and sexual reproduction

Mitosis is probably the simplest way in which a cell or organism may reproduce itself, but many unicellular organisms (e.g. *Paramoecium* which you saw in the TV programme of Unit 18), and most higher organisms, indulge in a more elaborate process which gives rise to a great variation between the genotypes of individuals within a population. This process

involves the combination of genetic material from two individuals, and is well known under the name of sexual reproduction. It is obvious that this type of reproduction will not produce offspring identical to one parent, but ones which are genetic mixtures of both the parents. However, the amount of variation produced is actually very much more than would be achieved by a simple blending of parental characters, and it is worthwhile examining the process in a little more detail.

In animals, specialized reproductive cells, called *gametes* (the egg and sperm cells) are produced in the *gonads* (called ovaries and testes) by each parent. These fuse, in the process of fertilization, to form a *zygote*, which has a single nucleus containing the genetic material from the two gametes. The zygote then begins to divide, mitotically, to grow into the new organism. In plants, gamete formation also occurs, but follows a rather different pattern.

The process by which the gametes are produced in the gonads is not mitosis, but a different form of cell division called *meiosis*. Clearly the gametes may have to be specialized in various different ways, for example they may have to be motile, like most sperm, or be able to withstand the rigours of the outside world for a considerable period of time. But one specialization they must all have in common is that they shall carry only half the normal (or *diploid*) chromosome number of the adult, and for this reason they are said to be *haploid*. Thus, in man, the eggs or sperm will have 24 chromosomes instead of 48 as there are in the adult; in *Drosophila* 4 instead of 8; in maize 10 instead of 20. The significance of this is obvious. If the gametes contained the full chromosome number, when they fused to form a zygote, the latter would have double the chromosome number of the parents, and so on at each generation. Not only would this not produce the genetic continuity stressed in Unit 17, but in a few generations the organism would soon consist of nothing but chromosomes.

Thus in one sense, meiosis is merely a specialized form of mitosis that reduces the chromosome number of the daughter cells by half, and does so in a very simple manner. In a diploid organism the chromosomes are in *homologous pairs*, that is to say there are two similar chromosomes in the nucleus, containing genes determining the same proteins (and thus phenotypic† characters), one inherited from each parent.

*Meiosis in* Drosophila

An example of a diploid organism much used in genetical research is the fruit-fly *Drosophila*, mentioned above. This animal is shown in film strip 19/20(a) and in Figure 8, p.21. It normally has eight chromosomes, four homologous pairs (Fig. 3).

You will recall from Unit 17 that, in simple cell division (mitosis), these chromosomes will line up across the spindle† in a single plane, and it can be shown that each chromosome has replicated into two identical chromatids, one of each being later drawn into each daughter cell. Thus, in the case of *Drosophila*, each daughter cell will have one chromatid from each of the eight chromosomes, making up the four homologous pairs. In the early stages of meiosis, on the other hand, the two members of each of the homologous pairs of chromosomes line up across the spindle, i.e. in the equivalent position occupied by the two chromatids of one chromosome in mitosis (Fig. 4, p. 18).

*Note: Figures 4–6 are highly stylized diagrams intended to clarify the steps in the process. In the film strip and appendix you will find a series of actual photomicrographs. These will show you the type of material from which we obtain the information used here.*

gamete

zygote

meiosis

haploid

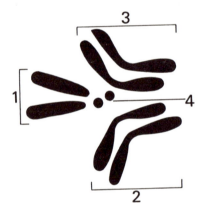

*Figure 3 The four homologous pairs of chromosomes in* Drosophila.

In each homologous pair, the chromosomes are associating very closely.

Figure 4 *The four homologous pairs of chromosomes arranged across the spindle at the start of the first meiotic division in* ♂ (male) Drosophila.

Consider just one pair for a moment. At this point both chromosomes can be seen to have replicated into two chromatids. The chromosomes consisting of pairs of identical chromatids, are drawn away from each other, one pair going to each side of the cell, just as the single chromatids are in mitosis. Thus, when the *first meiotic division* is complete, the two daughter cells each have only one chromosome (that is, two chromatids) from the original homologous pair (Fig. 5).

**first meiotic division**

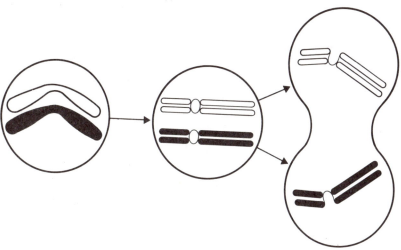

Figure 5 *First meiotic division, illustrating the behaviour of only one pair of homologous chromosomes in* ♂ Drosophila.

This is promptly followed by a division in each of the daughter cells which closely resembles a normal mitotic division although, of course, with only half the original number of chromosomes. The chromosomes lie singly across the spindle, and one chromatid from each is then drawn into each daughter cell. This completes the meiotic division, in which four sperm have been formed from the original cell, each with a chromatid from only one chromosome of each of the four homologous pairs (Fig. 6).

**second meiotic division**

As Figure 6 indicates, this process will give rise to four haploid sperm. Remember that one chromosome of a homologous pair is paternal in origin and one is maternal. So, if there were in fact just one pair of chromosomes involved, as in the figure, the sperm would contain genetic material either from the maternal or from the paternal side of the parent cell, but not mixtures of both, i.e. they would resemble genetically one or other of the two grandparents' gametes, which fused to form the genotype of the parental cell.

There are, however, *four* pairs of homologous chromosomes, so each sperm will contain *four* chromatids derived from those original eight chromosomes.

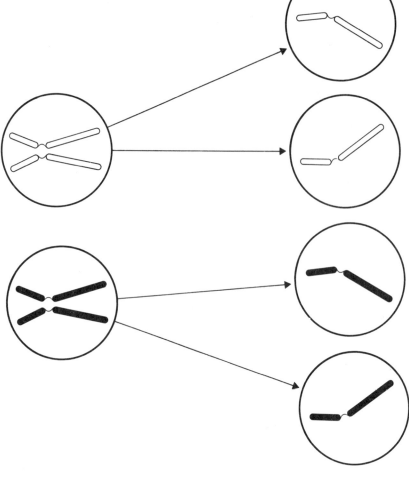

*Figure 6   Second meiotic division.*

**From the information given you here, can you calculate how many different genotypes (disregarding possible mutations, etc.) you might expect to find among the sperm arising from one cell by meiosis?**

No, you cannot.
If the homologous chromosomes all pair the same 'way up', relative to one another, across the spindle, then there will be only two different genotypes just as with the single pair of chromosomes. If they do not, then there will be a greater variety of genotypes.

To calculate how many different genotypes to expect in *Drosophila* sperm, you must first do part 1 of this Unit's Home Experiment. Just looking at photographs of the chromosomes of a cell during meiosis, as you can do in the film strip 19/20(a): 6, 7, 8, will not give you the answer. You cannot tell whether one of a pair of homologous chromosomes is of maternal or paternal origin from its appearance. It is therefore necessary to infer the answer to the above question by indirect means.

Unfortunately, it has not been possible to send you actual cultures of the organism we have chosen, so instead we grew them and photographed them in the Open University's laboratories. You can see in these photographs what you would have seen through the microscope.

**Now do part 1 of the Home Experiment**

From the photographs in part 1 of the Home Experiment, you will have deduced that it is purely a matter of chance which way up the homologous pairs lie across the spindle.

19

You can therefore now answer the question: 'how many different genotypes will be represented in the sperm of *Drosophila*?' If you have any difficulty, you can draw up a small table.

See Answer 1, p. 76.

An example of the actual effect this kind of redistribution may have on the genotypes carried by the sperm is given in the answer to the next question.

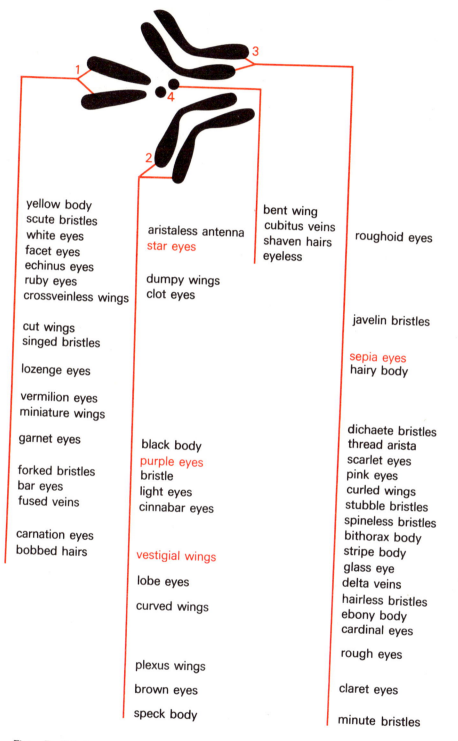

*Figure 7    Relative positions of some of the genes carried on the chromosomes of* Drosophila melanogaster.

Figure 7 shows the relative positions of some of the more important genes on the chromosomes of *Drosophila melanogaster*. (You are not required to memorize any of these; the figure is given for illustration only.)

20

In the Home Experiment, the orientation of the chromosomes could only be seen to determine the sequence of spores in the ascus of *Sordaria*. However, using the information in Figure 7, we can see some of the more important genetic effects that this type of redistribution may have. For example, on chromosome 2 there appears the gene for 'vestigial wings'. This is a recessive character, so that if the adult shows it, it means that the gene must appear in both the homologous chromosomes of pair no. 2. It is obvious then that it will appear in *all* the sperm of a male with vestigial wings. If, however, the adult has normal wings, it may be that it has one 'normal' gene and one 'vestigial' gene, the normal one being dominant. In this case, half the sperm would be expected to carry the normal gene and half the vestigial. If we take the same situation with another recessive gene on a different chromosome, say the gene for 'sepia eyes' on chromosome no. 3, the same will apply.

*Figure 8* Drosophila *with 'normal' and 'vestigial' wings.*

**Thus, if we have a male with normal wings and normal eyes, carrying a recessive gene for vestigial wings and one for sepia eyes, how many different genotypes in the sperm would you expect with regard to these two characters?**

See Answer 2, p. 76.

In fact, it is clear that either of the genes for the one character is equally likely to find itself in a sperm with either of the genes for the other character, if it is carried on a different chromosome. It follows, therefore, that these two characters are being distributed, or 'assorted', independently of one another.

independent distribution of character

This will not, of course, be the case where two characters are carried on the *same* pair of chromosomes. For example, the recessive gene for 'purple eyes' is carried on the same pair (no. 2) as that for 'vestigial wings'. There can therefore only be *two* genotypes for these characters among the sperm. Precisely what they are will depend on whether the 'purple eyes' recessive is on the same member of the pair as the 'normal wing'.

*At this point you should do the second part of the Home Experiment, if you have not already done so.*

21

*Crossing over*

You will have seen from your examination of spore formation in *Sordaria* in the Home Experiment that in fact there is greater recombination of characters than can be explained merely by the random arrangement of the homologous chromosome pairs.

You have seen evidence that the two members of each pair may exchange whole lengths of material (part 3 of the Home Experiment) between the time they replicate into chromatids and the time they pull apart. This process, called 'crossing over' is very general indeed, and occurs during meiosis in the male and female of almost all other organisms, and the female of *Drosophila* (Fig. 9). The case we have been considering above is exceptional in that crossing over does *not* occur. Thus, if we consider the same two recessive characters ('vestigial wing' and 'purple eyes') during *egg* production, it can be shown that about 15 per cent of the gametes no longer show the parental association. If we take two genes which are further apart on the chromosome, say 'vestigial wing' and 'star eyes', this percentage rises, indicating that the further apart the two genes are on the chromosome, the more likely they are to get separated by crossing over.

crossing over

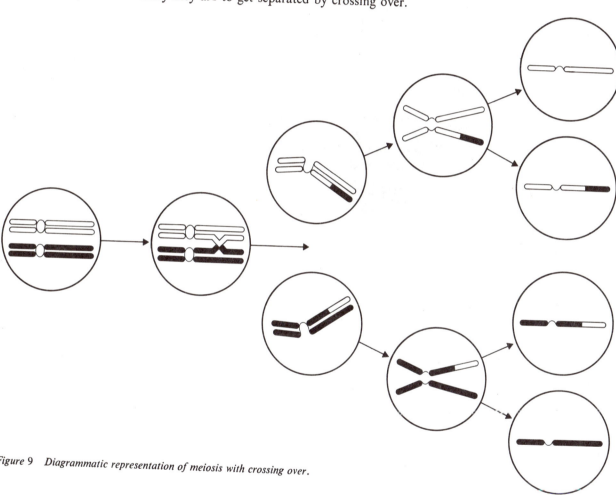

Figure 9  *Diagrammatic representation of meiosis with crossing over.*

Crossing over only occurs between two of the four chromatids at any one chiasma (see Home Experiment Notes, part 2). However, other chiasmata may form further along the same chromatid, or between the other two chromatids, forming double cross overs (Fig. 10). Sometimes, as you have seen, three out of the four chromatids may be involved, one crossing once with one opposite member and, further down, with the other.

The number of cross overs varies very much between species, and between different chromosomes of the same species. An average figure would be two or three chiasmata formed at each pairing of homologous chromosomes.

From this you can see that meiosis provides a very great source of genetic variation, partly due to the assortment of maternally and paternally derived chromosomes, caused by the random orientation of the homologous pairs before division, and partly because of the exchange between chromatids in crossing over.

## The extent of recombination

recombination

There is effectually no limit for the number of possible recombinations in the genotype that may be produced by meiosis. For example, a single human genotype probably carries in the region of 100 000 pairs of genes. If only twenty of those pairs showed differences, one would expect over a million different genotypes in the gametes due to assortment and recombination. In fact, it is likely that many more than twenty pairs of human genes show dissimilarities, so that in an average ejaculate of 200 million sperm it would not be expected that any two of them would share an identical genotype. The same variation will of course apply to egg genotypes, so the probability of any two children of the same parents inheriting the same genotype are quite negligible. Indeed, it is highly improbable that any two human beings who have ever lived will have had the same genotype (except where the zygote has split into two *after* fertilization, giving identical twins).

The effects of mutation in producing entirely new genes must be remembered too, as these will also be undergoing recombination. Probably 20 per cent of the people around you contain some mutation which arose for the first time in either their mother or their father—and this figure excludes dominant lethal mutations, as you will not have seen these people!

There are other factors giving rise to genetic variation. One of these is the production of individuals with more than the normal set of chromosomes. These individuals are called *polyploids*, and though they seldom survive if they are animals, many important crop plants are in fact giant polyploid versions of a normal ancestor. You will find more about these in black-page Appendix 2.

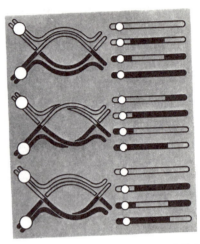

*Figure* 10 *Diagrammatic representation of various combinations of cross overs between different chromatids.*

polyploid

## 19.3  Genotype and Phenotype

These terms were defined in Unit 17. The variations we have been discussing above are variations in the genotype of the organisms, i.e. the genetic composition of the individuals. These variations will show as changes of some kind in the phenotype, unless they involve wholly recessive genes.

In instances such as the gene for 'vestigial wing' in *Drosophila* the animal either shows the abnormality or it does not. But in other cases a single recessive gene may in fact 'show through', so to speak, even if only to a very small extent. One example is the recessive gene for Haemoglobin S, mentioned in section 19.2.1 above. As you will see in 19.6.4, it is possible to observe the effect in the heterozygote, although only under certain conditions. Thus it would be rash to assume that a single recessive gene never has an effect of evolutionary importance on the phenotype.

During an individual's lifetime, phenotypic changes will occur continuously, an obvious example being growth, maturation and ageing, but changes to the genotype will be small (random mutation) or absent. Furthermore, changes in the genotype will not be heritable if they occur in any of the great majority of the organism's cells—its body cells—but only if they happen to occur in sperm or egg mother-cells in the gonads. Thus, within the life of an individual organism, we expect radical changes to occur in the phenotype, but probably not in the genotype. Nevertheless, the nature of the genotype may profoundly affect not only the original structure and function of the phenotype but also the phenotypic changes which occur.

For example, it is a matter of common experience that use (= 'training') is profoundly important in the development of the body for various forms of athletics. For many forms of sport, one of the limiting factors on performance is the efficiency of the circulation in getting oxygen to the muscles (see Unit 18) and in removing lactic acid as it accumulates. A great deal of the difference in stamina between, let us say, an athlete and the authors is, in fact, merely a reflection of the relative capacities of our circulatory systems.

*use affects phenotype*

Use not only increases the size of the appropriate muscles, but also develops the blood vessels supplying them. It also greatly increases the amount of blood circulated per minute, sometimes the crucial factor. This is achieved, rather surprisingly perhaps, by increasing the capacity of each heartbeat—i.e. the blood pumped per beat—at all times. The athlete and the authors, when at rest, require about the same amount of blood to be pumped round their bodies, say 5 litres per minute. This will be achieved by the heart beating about 75 times per minute in the authors, but only 60 times per minute in the athlete. In a fierce race, the more blood that can be got to the muscles the better. The authors' heart rates will rapidly rise to 180 per minute, at which rate the amount of blood pumped per beat is falling rapidly, as the heart does not have time between beats to relax and fill properly. Thus this rate, perhaps the maximum, will be reflected in a maximum output of about 25 litres per minute. At this output, however, the athlete's heart is ticking over at a mere 100–120 beats per minute and when pushed harder, it can give an output of 35 litres per minute.

Training would improve the performance of the authors' hearts to give a larger, slower beat, but would never enable them to compete with an athlete. The evidence is that to be able to reach the really high output

figure, it is necessary to start with a slow strong beat. The characteristics of the heart will initially be determined by the genotype, and this will affect what changes can be induced in the phenotype by training. The same training will not therefore produce equally good results on two genotypes. Whereas the athlete may be able to pass such potential on to his children, it is unlikely that the authors will to theirs.

This relationship between genotype and phenotype has been a matter for much thought over the years. The Danish botanist Johannsen started an experiment in 1909 in an attempt to distinguish which variations in a variety of garden beans were due to genetic differences and which were phenotypic. The variation he was measuring was in the size and weight of the bean seed itself.

He used a commercial 'variety', which in reality consisted of a number of different genotypes. There was a wide range in the sizes of the beans, and he separated out the smallest and the largest ones. He grew them under similar conditions, and when the plants flowered he self-fertilized them. (That is to say he fertilized the female part of a flower, the *ovules*, with pollen from other flowers on the *same* plant.) This ensured that no new genotype was introduced, only that already present in the original seed from which the plant was grown.

We may call the plants grown from the small seeds group A, those from the large seeds, group B. After each plant had been self-fertilized, the beans grew and ripened in the normal way. He then collected them and measured them.

> **Would you expect that the average size of the beans produced by the plants of group A were:**
>
> **(a) the same as the average size of those produced by group B (large seeds)?**
>
> **(b) smaller than those from group B?**
>
> See Answer 3, p. 76.

There was a considerable size range within the progeny of both groups. Johannsen then took the largest and the smallest beans from *within* each group, grew them, self-fertilized the flowers and collected the beans.

He found that the average size of the beans grown from the smallest beans of the group A line was the same as the average size of those grown from the largest of the same group. Thus within the group his selection was not effective.

> **Why not?**
>
> See Answer 4, p. 76.

Thus group A was 'breeding true' for smaller beans, subject to the normal variations expected from the effects of the environment and experimental errors.

The same applied to group B.

> **Would you expect that the average size of the group B line remained greater than that of the group A line?**
>
> See Answer 5, p. 77.

Many other instances of the effects of the environment on phenotypes spring to mind. For example, a young couple may spend their summers on the Costa Brava and their winters skiing—and be elegantly suntanned all the year round. However, they may eventually have children and cannot

25

afford to go on holiday. Not only will they lose their tan—but their baby will be born without a trace of it.

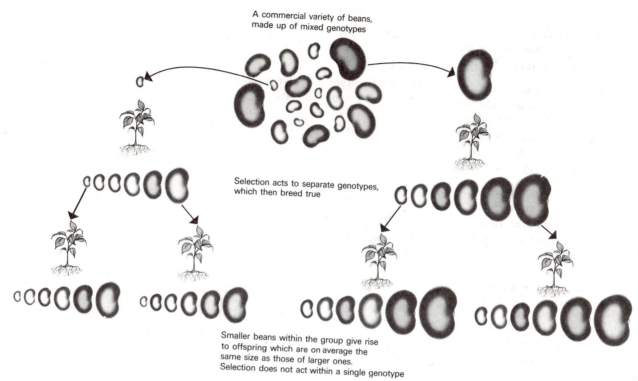

A commercial variety of beans, made up of mixed genotypes

Selection acts to separate genotypes, which then breed true

Smaller beans within the group give rise to offspring which are on average the same size as those of larger ones. Selection does not act within a single genotype

*Figure 11 Diagram summarizing Johannsen's experiment.*

*Obviously, the differences in size of the beans possessing a single genotype or in a 'pure line' were caused by variations in the availability of such things as light, dissolved salts, and perhaps the physical nature of the soil encountered by the roots of different plants. The position of the bean in the pod could also have made a difference.*

Various species of the garden and pot plant, Hydrangea, show colour variations of their flowers. Thus it is possible to cultivate generations of Hydrangeas with blue flowers, only to discover that if transferred to a soil deficient in iron they become a pale pink. If you want to turn your pink Hydrangea blue, just water it with iron salts; but you will also have to water its progeny, as the change is not heritable.

We have so far considered the effects of the genotype on the phenotype at a molecular level (one gene = one polypeptide—in Unit 17), and also a little of the effects of the environment on the phenotype. What of the effects of the phenotype and the environment on the genotype? Traditionally, the environment was assumed to affect what we would now call the genotype through the phenotype in a very direct manner. Thus the environment, in the form of the crocodile, pulled out the nose of the Elephant's child (the phenotype) and thereafter ( = the genotype) elephants had long noses.

Can you in fact produce a plausible explanation of how this could come about? Many people have, including Charles Darwin, but we will deal with this aspect later (section 19.5). Modern evidence does not support the idea that changes in the phenotype can act on the genes to produce a corresponding change, i.e. one that could imprint that change on the genotype. This must be clear from what has been said both in this Unit and Unit 17. But this is not to say that the phenotype has *no* effect on the genotype. It does, after all, provide the environment for the genes, and thus may affect them. For example, you will see from black-page Appendix 1 that, if the oxygen level in the tissues drops, mutations of the genotype due to certain causes are inhibited. On the other hand, if the phenotype *Drosophila* is made to live with a higher body temperature, the mutation rate rises.

**effect of the environment on the genotype**

26

The external environment in which the whole phenotype lives, also affects the genotype both directly (radiation causing mutation) and indirectly, through the fate of the phenotypic expressions of the genes, that is the individual organisms.

### 3.1 Summary of sections 19.2 and 19.3

1   Although we have seen from Unit 17 that cell multiplication (mitosis) involves a basically conservative mechanism, variation does occur. It occurs by mutation, which is a sudden change in the arrangement of the nucleotides in the DNA of the chromosomes. This results in changes in the polypeptides determined by the structure of the DNA, and so in detectable changes in the cell or organism. Mutations appear to be essentially random, and may occur to any gene at any time, though some genes appear more susceptible than others. Because of the complexity of living organisms, mutations are more likely to be deleterious than beneficial, though when they are recessive they may not affect the phenotype for many generations.

2   Variation in the genotypes of organisms that reproduce sexually will arise because:

(a) each is an equal mixture of material from two parents;

(b) the gametes carrying the material from the parents are produced by meiosis. Meiosis itself involves the random assortment of whole chromosomes as well as genetic exchange between parts of them. Thus almost limitless recombinations of the existing genes may take place;

(c) mutations occur in the cells giving rise to the gametes, as in any other cell.

3   All these variations are heritable, but there will also be phenotypic variations which are not. It may be important to distinguish between them.

**Section 4**

## 19.4  Populations

Until now we have considered genetic variation mainly as it may arise in individuals, by mutation, assortment and recombination. So far as the reality of genetic change within any kind of sexually reproducing organism is concerned, certainly from an evolutionary point of view, we have really only been considering models—or at least prototypes. Most organisms do not live as isolated individuals, and even if they did genetic change in such individuals would be hard to locate. But neither do organisms really live together as a species (a term we will examine later) or 'kind'—although we tend to group them as if they did. For example, the genotype of one lion is of little significance to the lion species as a whole, but to try to consider change in the genotype of the whole species becomes impractical, considering that lions may live thousands of miles apart, separated by rivers, mountains and deserts, with no direct communication between them.

If we are to consider genetic change in relation to possible evolutionary mechanisms, with what unit are we involved? Essentially it is a geographical one, though the geography may not always be obvious. It is that group of a particular kind of organism the members of which encounter one another sufficiently often and in such a manner as to result in their interbreeding regularly. Such a group is termed an interbreeding population, or just a *population* for short, and ideally every member of it has an equal chance of meeting and mating with every other appropriate member, but very much less chance of mating with a member of another population.

population

This ideal population may be achieved if the group is confined in a relatively small space with definite boundaries—say rabbits on a small island. More often than not, however, it is more a question of degree. For example, there is almost continuous coverage of the British Isles by the Brown Rat. Even so, we do not normally think of them as providing a single freely interbreeding population, though if we viewed them in terms of the evolution of the whole species over some millions of years we might. But a rat from Glasgow will not often requite its love with one from Tunbridge Wells. For purposes of detailed study by members of the Open University, it would be much more usual to consider the rat population of the village of Milton Keynes and a one-mile radius around it, bearing in mind the possible influences on it of the Bletchley rats, four miles away. All the same, a longer-term study, say over ten years, would require allowance to be made for more mixing and movement, and the net might have to be cast wider. You will meet such an example in section 19.6.1, in the spread of resistance to the poison Warfarin among the rats of Western Britain.

So the boundaries of these population units are often arbitrarily drawn, depending on the time scale and the size and mobility of the individual organisms involved, and on where one particular type of environment may be said to end and another begin. Nevertheless, these populations do seem to form the units of evolution and they have a real genetic significance as you will see in section 19.6. Of course, any such population may really only exist for a limited period in time before its edges become blurred.

The main difference between the colloquial sense of the word 'population' and the genetic one is the clear implication in the latter that the members interbreed freely and randomly. Thus two ants' nests close together in a garden would not form a single population from a geneticist's point of view, because they normally breed only within their own colony (and then only a select few are involved). To the average householder, however, it would seem fair to say that his garden had a large population of ants.

Thus a population in the ordinary sense of the word is simply a geographic entity, and can easily be rather larger than the geneticist's one. For example, even today one could not truthfully say that the 'population of London' was a freely interbreeding unit. It is not really true to say that your chances of meeting and marrying the boss's daughter from Primrose Hill are as good as those of marrying your next door neighbour's in Camden Town. It most certainly is not so if either you, or she, is a coloured Moslem from Southall. Thus genetically speaking London is a number of separate populations—perhaps to the detriment of all.

## 19.5 Evolutionary Theories

In the introduction, we gave our reasons for making the assumption that species had developed by some process of evolution, rather than by any process of special creation. It is clear that such evolution cannot take place unless change can occur in the genotypes. It is this change and some of the causes of it that we have been examining so far. But, unless there is also a great deal of genetic stability or continuity (Unit 17), it is equally impossible to visualize any evolutionary process. If organisms came out quite differently at each generation, no character could be said to be heritable and any idea of a progression to a new or different 'kind', which is what is implied by the idea of evolution, would be meaningless.

So what we are considering is a fairly balanced system, basically stable but with an ever-present element of variation; the units of this system, and those of evolution itself, are the interbreeding populations discussed above. No case for any evolutionary mechanism—or for that matter for its occurrence at all—can be convincing unless it can be shown that it can account for a lasting and directional change in such a population. Indeed, it is far more convincing if the postulated mechanism does not merely accord with the changes, but can be shown to produce them, though clearly even this is not proof positive of its universal applicability.

You must remember that the arguments over the nature and causes of evolution have been taking place for many years, but, until recently, without the genetic information you now possess. The benefits of hindsight are considerable when looking back on the early theories from the vantage point of the 1970s.

Even so, it is quite surprising that three or four hundred years ago, when almost no one would have even considered the heresy that one type of animal could have evolved from another, there was general acceptance of the idea of the inheritance of characters acquired during the life of the phenotype. Ancient folk tales are to be found from all over the world on the general theme of the Kipling story of the Elephant's child, mentioned above. For example, hares are supposed to have split lips as a consequence of the uncontrolled laughter of a bygone Chinese hare.

The implication that generations of earlier hares had intact lips—and that if lips could change, so could ears, fur, size or anything else—seems to have worried no one.

However, by the end of the eighteenth century, a good many thinkers had come to the conclusion that there *had* been organic evolution, though it had not been formulated as a scientific theory, nor had any convincing mechanism been put forward to account for it. The idea was not acceptable to the public at large and was in direct conflict with the teachings of the Christian Church. However, the power of the Church to control philosophy and discussion was on the wane, and it is probably no accident that it was at this point in history that an evolutionary interpretation came to be put on facts which had been under consideration by natural philosophers since Aristotle's time (384–322 BC). Scientists, like others, tend to operate within the social climate of their time. Belief in the immutability of a way of life and the idea that nature has already been explained, is a considerable encouragement to the belief that species are also immutable. Buffon (1707–1778) produced what was probably the first clear statement of belief in organic evolution, and he did so working in Paris in the climate

which led to the French Revolution, a time when many sacred cows were slain. (See *The Roots of Present-Day Science*.)

Erasmus Darwin (1731–1802), grandfather of Charles Darwin, wrote as an evolutionist who believed that all life—including, by implication, human life—originated from a common source. Whereas Buffon was clear that the mechanism was the inheritance of acquired characteristics (see below), Erasmus Darwin avoided coming to grips with this part of the problem. Lamarck (1774–1829) was a brilliant biologist with a very wide knowledge of both plants and animals. He was much influenced by Buffon's thinking—indeed he tutored Buffon's son—and he formulated the first really clear and comprehensive evolutionary theory.

Like Buffon, he believed firmly that the mechanism was the inheritance of acquired characters, so that today as a 'doctrine' it still bears his name —Lamarckism. The problem encountered by all these men, and some of those who followed them, was the need to explain how variation could occur and the variants be sustained and perpetuated. Common sense suggested that offspring were a blend of their parents, and the blending was usually held to be one of bloods. However, such a system would produce increasing uniformity, and a particular character would become progressively diluted at each generation. Thus, the idea that phenotypic changes are passed on (in the blood) was not only plausible, it also provided an explanation of how new characters could be sustained and improved. If, for example, both parent giraffes somewhat increase the length of their necks reaching up into the leaf canopy of trees, their offspring will have longer necks. The parents are sharing the same environment, so *both* their necks are affected—reducing dilution. The offspring are again confronted with the same environment, so the effect is progressive. Lamarck was quite clear that it was the action of the environment on the organisms which produced the change, rather than any vague accident or unspecified design of the Creator. Many of his views on the interaction of the organism and the environment, and on the nature of the evolutionary process generally, were close to those held by a majority today. His importance in the development of modern evolutionary theory is often underestimated. He did not in fact adopt the very naïve views sometimes called Lamarckist. He quite specifically rejected the idea that crude direct effects of the environment on an organism would be transmitted in any way to its offspring. Thus, he would not have expected that cutting the tails off the adults of generations of mice would result in the development of a tail-less breed of mouse. Yet this experiment has been done at length in an attempt to disprove 'Lamarckism'. In Lamarck's view, the heritable element was the change produced in the phenotype by the phenotype itself, by use or disuse. The whale's flippers or the kangaroo's forelimbs could be considered classic examples of the results of disuse in this context.

One of his main opponents in this evolutionary thinking was Cuvier, whose theory of catastrophism was mentioned in the introduction. But Cuvier's views were largely demolished by the work of the British geologist Sir Charles Lyell (1797–1875) (see Unit 26). He provided convincing evidence that geological evolution was a continuous, and continuing, process rather than a series of Creator-induced catastrophes.

Thus, during the latter part of the eighteenth century and the early part of the nineteenth, scientific thinking was largely along evolutionary lines. However, the only mechanism which had been advanced and supported in any detail to account for biological evolution was that of the inheritance of acquired characters, and their fixation in the species by the action of the environment.

This remained substantially true until 1858. That year saw the joint publication of a paper, based on two separate pieces of work, by Charles

inheritance of acquired characteristics

30

Darwin (1809–1882) grandson of Erasmus, and Alfred Russell Wallace (1823–1913). In this paper, which was read to the Linnean Society of London, the authors suggested that the main factor producing evolutionary change was what they called *Natural Selection*.

**natural selection**

Their argument, amplified a year later by Darwin in *The Origin of Species*, was essentially this: organisms tend to produce more offspring than their environment will support, therefore a large number will perish before completing their reproductive lives. Those whose phenotypes are better suited to their immediate environment will, in the long run, have a greater chance of being among the survivors than the less well suited. This in turn will mean that over a period of time the better suited phenotypes will predominate in the population.

The ingredients of this process are (a) *overproduction* of the organisms in terms of what its environment will support thus leading to (b) *competition* among (c) heritable *variations* of the phenotype.

**overproduction**
**competition**
**variation**

There is no doubt that these ingredients are normally present within any natural environment. Overproduction is quite evident. A plaice lays half a million eggs at a single spawning, a cod several million and an oyster over 100 million. A pair of rats may produce an average of ten young in a litter, three or four times a year. The offspring themselves will mature in three or four months, with the result that a pair of rats in an ideal environment could be one of five hundred pairs within a year.

Bacteria dividing every thirty minutes could soon engulf the Earth's surface, and if every one of the 700 000 000 000 spores of the puffball fungus were successful, one such could in theory give rise to a mass of puffballs greater than the Earth in eighteen months or so. Common experience shows that all the organisms we meet have a much greater reproductive potential than they are likely to be able to achieve for long. Man is in the interesting position of being able, so far, to 'bend' the environment continuously to keep up with almost unrestrained population growth, though even this growth is less than he is theoretically capable of.

**competition from overproduction**

Given the first ingredient, the second must almost inevitably follow. Taken at its simplest, if there are 100 animals born into an environment which only provides enough food for twenty-five to grow and mature, there will be great pressure on all the individuals and they will be directly in competition with one another. As the food supply becomes heavily taxed, they may all become undernourished. At an individual level it may be pure chance which one succumbs and which survives. One which has been more successful in feeding itself and is thus more active and stronger may nevertheless be caught by a predator, but all the same its chances of survival are better; in the long run the more successful food gatherers would be expected to survive to maturity more commonly than the less successful. Clearly the third ingredient is involved at this point. If all the 100 individuals are identical, it will only be pure chance that determines the survivors. If they are genetically identical but show phenotypic differences, any advantage conferred upon an individual by such a difference would only be passed on to its progeny if there is in fact some direct effect of phenotypic change on the constitution of the germ cell, which most geneticists would deny. In fact, the animal would be in the position of Johannsen's beans belonging to a pure line—even when there was 100 per cent selection of the larger beans within the line, the beans produced by these plants were no bigger on average (p. 25). This experiment, together with what you have seen of modern genetical theory and research, make it almost certain that Darwin's 'natural selection' will only operate to produce lasting change in a population if there is *genetic* variation within that population.

When Darwin and Wallace were formulating their theories, however, they did not know this. Although the first steps in genetics were taken by an

Austrian monk named Gregor Mendel and published in 1865, during the time that Darwin was actively improving and elaborating his theory, he never had the benefit of them. Mendel published his results in a single paper in the Brno Natural History magazine, not a widely read journal, and indeed they remained unnoticed until the basic 'rules' of genetics were rediscovered, as was the paper, in 1900, some time after Darwin's death. Had Darwin known of them, it probably would have made a great difference to the development of his theory. He always felt that the weakest point of his argument lay in the absence of an explanation of how variation was transmitted. Not having any of your knowledge of genetics, he never believed the observable mutations in domestic animals ('sports', as they were called) to be a major factor in providing the variation on which natural selection could work. He assumed that such changes were bound to become more and more diluted at each generation, in a form of 'blending inheritance'.

Darwin is commonly held to have produced 'the' theory of evolution, and to have accounted for it by the idea of the 'survival of the fittest'. But neither of these statements is really true. As we said above, the idea of biological evolution was a topic of much discussion among scientists and natural philosophers during the latter half of the preceding century. Darwin's great contribution was to have suggested, in his hypothesis of Natural Selection, a practical and easily understandable means by which evolution could have taken place.

What Darwin did was to put the idea of organic evolution into the framework of a real scientific theory and to present an enormous amount of careful observation and reasoned argument to support it. He himself accepted evolution as fact at the start of his work; his main theme was to account for adaption and to explain evolution in terms of it. The idea that natural selection was the agent by which the environment affected species (and indeed that this could lead to entirely new species) was entirely original and of the greatest importance. It is essentially the view held by most biologists today. But Darwin believed that natural selection was merely the most important of four major factors. The next most important was the inherited effects of use and disuse (i.e. 'Lamarckism'); then came the inherited effects of the direct action of the environment on the phenotype (an occurrence Lamarck did not accept, but one adopted by his followers or 'neo-Lamarckists'); and finally the occurrence of 'sports' or mutants. The condensation of the idea of natural selection into the phrase 'survival of the fittest' was not Darwin's, but an enthusiastic follower's, Herbert Spencer. It is rather misleading, suggesting that the process produces the very best in some absolute sense, rather than simply tending to eliminate more of those members of a population less well adapted to a particular environment. However, it was u phrase that caught the imagination, particularly of those interested in its application to human society and ethics.

Thus Lamarck and Darwin were really not so very far apart in their thinking, though this is not to belittle the extent and accuracy of Darwin's work or the overwhelming importance of the idea of natural selection. It is rather ironic therefore to find that bitter controversy raged throughout the end of the nineteenth and the beginning of the twentieth century between 'neo-Lamarckists', who postulated direct environmental effects that Lamarck did not believe in, and the 'neo-Darwinists', who interpreted natural selection in ways Darwin would probably have laughed at.

Darwinism

### 9.5.1 Can these theories be verified?

It is, of course, nearly impossible to *disprove* either theory, or for that matter to *prove* that any such theory has accounted for all evolution; it is not enough, even today, to say that our knowledge of genetics shows that Lamarck was wrong. It makes it likely that he was wrong because it is hard to think of a Lamarckian mechanism compatible with what we know (unless, of course, *you* did on page 26). But what we *can* do is to make predictions according to either theory, and then test the predictions. This is not very easy to do for several reasons, one of which is the time-scale involved.

You have, however, already met one example where the generation time of the organism is as little as thirty minutes and evolutionary processes which would take thousands of years in a mammal could perhaps be telescoped into months. This example is *E. coli*, the bacterium cited when we considered mutations arising in populations of it in culture (p. 11). These mutations conferred a resistance to the antibiotic streptomycin, in one case, and to attack the T4 phage, in another. Exposure of a population of *E. coli* to these agents represents a fundamental change in its environment, in the first case by subjecting it to a massive attack by a parasite not previously present in its environment, and in the second by saturating the environment with a chemical that upsets the replication process in the normal bacterium. Both of these are rather acute examples of the sort of change which could be encountered by any natural population.

selection or induction

Not all the individuals were killed; in both cases some survived and formed new colonies, the genetic constitutions of which were different from those in the original population. In the case of phage resistance, the experimental technique used is as follows. The bacteria are grown in a liquid medium, let us say in ten tubes, each tube being seeded with one or a small number from a single clone. These are then incubated until the bacteria have multiplied to a density of about $5 \times 10^9$ per cm³. Whilst the bacteria are being incubated, ten Petri dishes are prepared with a nutrient jelly containing the bacteriophage in very large numbers, much in excess of those of the bacteria. The bacterial cultures are then poured onto the jelly containing the phage and the dishes are incubated. Sensitive bacteria are destroyed in a few minutes. Any resistant ones will grow into visible colonies within twelve to sixteen hours.

As we said above, these resistant ones must have a different genotype from the normal population, which is sensitive.

**How would you account for this change in Lamarckian or neo-Lamarckian terms?**

See Answer 6, p. 77.

**How would you account for them in terms of modern 'Darwinian' theory?**

See Answer 7, p. 77.

**In terms of these two theories, can you make predictions as to what the outcome of the experiment should be, numerically speaking?**

See Answer 8, p. 77.

Table 2 gives the results actually obtained from five experiments performed in the manner described above. Each figure represents the number of resistant colonies in each of the ten dishes (in the case of experiment 5 only nine dishes were used).

**Table 2  Number of resistant colonies in experiments with *E. coli***

|  | Experiment No. | | | | |
| Dish No. | 1 | 2 | 3 | 4 | 5 |
| --- | --- | --- | --- | --- | --- |
| 1 | 30 | 6 | 1 | 1 | 10 |
| 2 | 10 | 5 | 0 | 0 | 18 |
| 3 | 40 | 10 | 0 | 0 | 125 |
| 4 | 45 | 8 | 0 | 7 | 10 |
| 5 | 183 | 24 | 0 | 0 | 14 |
| 6 | 12 | 13 | 5 | 303 | 27 |
| 7 | 173 | 165 | 0 | 0 | 3 |
| 8 | 23 | 15 | 5 | 0 | 17 |
| 9 | 57 | 6 | 0 | 3 | 17 |
| 10 | 51 | 10 | 107 | 48 | — |

Do these findings agree with the Lamarckian prediction? Clearly they do not. They do, on the other hand, allow the 'Darwinian' view to be held. The distribution of resistant colonies is in fact what would be expected on statistical grounds from a spontaneous mutation rate between $1.1 \times 10^{-8}$ and $4.1 \times 10^{-8}$.

Thus we have firm evidence of an instance where a major change in a population has been brought about by selection of a genetic character which had already arisen in the population; furthermore, it is an instance where the so-called 'Lamarckist' prediction is not fulfilled. A much simpler and cruder way to test the 'neo-Lamarckian' theory, that is the direct effect of the environment on the genotype, is to perform some operation on many generations of adults, to see if the change begins to appear in the offspring. This poses many problems, not least the question of *when* the phenotypic change is expected to affect the genotype. After all, it is known that in the female mammal most, if not all, the egg cells are present in the ovary even before the female is born, therefore one could argue that the phenotypic change might have to occur before birth. Nevertheless, enthusiastic biologists cut the tails off hundreds of generations of mice without producing the forecast strain of 'manx' mouse. Dog breeders have been docking the ears (though no longer in Britain) and tails of Boxers for a great many years, but the puppies are still born with long ones.

*Figure* 12  *Parent Boxer dogs with docked ears and tails and their normal puppies.*

## 19.6 Micro-Evolution

We have established that the environment produces change in a population by favouring some heritable variations within it more than others. We have said that the favoured organisms are the more 'fit'. Used in the evolutionary sense, *'fitness'* does not mean quite the same as in the athletic sense. Observation may suggest that an animal is better fitted to a particular environment than others of its kind because it is better concealed, or because it is bigger, or because it seems more intelligent. This is really trying to put a value-judgement on a character in a situation which is so complicated as to make it meaningless. The proof of the pudding is in the breeding, which is to say that the only meaningful way to judge fitness is in relative numbers. If a variety within a population is common, it is reasonable to say it is well fitted to its environment. If it appears to be increasing or have grown at the expense of others, it is also reasonable to say it is better suited, or more fit, than they are. All the many factors really boil down to this one end factor, which can usually be measured. Thus evolutionary fitness is really reproductive fitness. An observer may form **reproductive fitness** an opinion as to *why* a variation is more (or less) successful, but the only way he can measure it is in terms of survivors over several generations. We can use an imaginary example of how selection may be observed to determine 'fitness' in the wild. Consider a population of mice, equal in all observable respects (a statement one would not dare make in a real example) except that half are light coloured and half are dark coloured. If the main predators are a pair of owls, higher losses are to be expected among the lighter variations, as they are more conspicuous at night. It is reasonable to assume that each pair of mice will average eight offspring per litter. We can start with a population of 200 new-born mice:

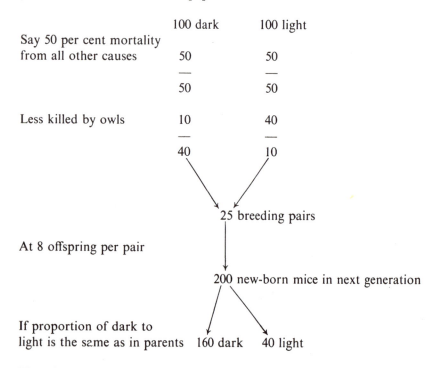

|  | 100 dark | 100 light |
|---|---|---|
| Say 50 per cent mortality from all other causes | 50 | 50 |
|  | — | — |
|  | 50 | 50 |
| Less killed by owls | 10 | 40 |
|  | — | — |
|  | 40 | 10 |

25 breeding pairs

At 8 offspring per pair

200 new-born mice in next generation

If proportion of dark to light is the same as in parents    160 dark    40 light

Thus the owls will soon make the light variations a rarity. We can put a figure to this 'fitness'. We can say that of the original 100 dark mice, 40 survived to breed and took a part in the production of $40 \times 8 = 320$ offspring. To maintain a stable population where there are equal numbers

of males and females, the average production of offspring must be two per pair. Assuming that this was the only breeding of our original mice, we can say that they averaged 3.2 offspring each rather than 2 or $\frac{3.2}{2}$, giving them a 'fitness' of 1.6. By the same calculation the fitness of the light mice under these circumstances was 0.4.

A real example is demonstrated in the TV programme of this Unit, when we deal with the case of the mutant form of the Peppered Moth. You will notice there that we do not refer to fitness as a ratio, but we mention simply the 'survival advantage' of one form over another as a percentage. This is because we do not have all the figures of the reproductive sequence given in the imaginary case, we are merely measuring a relative adult survival.

In the case of our imaginary mice, natural selection was producing a minor change in the population; within a few years it would have made light coloured mice a rarity. They would probably not have vanished altogether, for various reasons—new ones might have arisen by mutation, or migrated in from areas where the predators were not owls or where their surroundings were lighter, thus giving them a better chance.

Small but important changes due to selection can be shown to be occurring in real populations, and these small evolutionary steps are often said to be examples of micro-evolution, to distinguish them from major evolutionary steps such as the production of clearly distinct new species, such as we shall see in section 19.7. The change occurring in the population of the Peppered Moth, mentioned above, is an example of micro-evolution.

Many of the processes of adaption occurring as a result of selection pressures are not of merely scientific interest, but of the greatest immediate importance in our everyday lives. One common example of this is *adaptive immunity*, the acquisition by a population of immunity to agents, biological or chemical, which previously were severely damaging.

### 19.6.1 Adaptive immunity

You have seen how strains of *E. coli* resistant to antibiotics can be isolated in the laboratory. The same process, unfortunately, may occur within human communities to much more harmful bacteria (remember the radio programme of Unit 10). *Staphylococcus aureus* is a common bacterium found on the skin, in putrefying matter and elsewhere. If it gets into the tissues it is pathogenic, that is to say, it may multiply faster than the body's defences can destroy it, attacking the tissues and producing poisonous waste products. Depending on where it happens to be, it may be the causative organism of a sore throat, boils or a fatal blood poisoning. This bacterium, in common with other staphylococci, is generally very sensitive to penicillin, even in quite small doses. However, when a dose of penicillin just large enough to 'cure' the disease is used, a number of resistant individuals able to protect themselves by producing a penicillinase, are likely to be isolated—as in the culture of *E. coli*. These will not necessarily succeed in multiplying into large numbers (causing a relapse in the patient), because the normal bodily defence mechanisms may keep them in check. Nevertheless, they will be there in small numbers, perhaps in the discharge of a healing boil or the saliva of a recent victim of a septic throat. Eventually, one such resistant bacterium may infect another victim, and it will be found that the disease does not respond to treatment with the normal dose of penicillin. Usually a very much larger dose will still work, and so it will be used. Resistance to very large doses requires a further mutation, in addition to the first one. But once again the selection

bacterial resistance to antibiotics

36

process may take place, this time producing bacteria resistant to very large doses, often as large as it is safe to prescribe. It is in this way that modern medicine produced the 'Hospital Staph', a very tough organism, with the result that the level of post-operative infection—which had fallen very low indeed in the early days of antibiotics—has risen considerably. Usually the causative organism is an antibiotic-resistant one.

The first signs of staphylococcal resistance appeared soon after penicillin became extensively used in hospitals, as Table 3 shows.

**Table 3   Incidence of penicillin-resistant infection at a general hospital**

| Date | Total patients | Patients with penicillin-resistant strains |
|---|---|---|
| Apr–Nov, 1946 | 99 | 14 |
| Feb–June, 1947 | 100 | 38 |
| Feb–June, 1948 | 100 | 59 |

By 1950, a majority of staphylococcal infections in *all* British general hospitals were penicillin-resistant.

Resistance has subsequently appeared in certain staphylococci to all the major antibiotics, often as a triple resistance, that is the same strain being resistant to penicillin, tetracycline and streptomycin. Fortunately, not all the strains of *S. auraeus* have produced such a dangerous triple immunity.

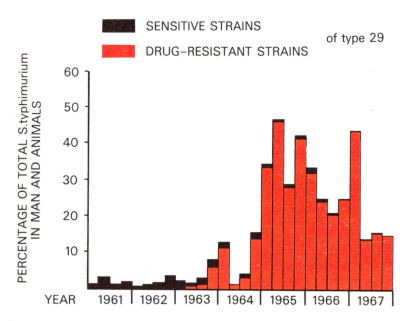

*Figure* 13   *Incidence of drug-resistant strains of the type 29 bacterium causing typhoid in man and animals, expressed as a percentage of the total number of diagnosed infections.*

It is also interesting to note that the much more dangerous bacterium *Streptococcus*, which was producing strains resistant to treatment with sulphonamides† (in the days before antibiotics were used), has fortunately failed to produce a penicillin-resistant strain. There are, however, mutant strains resistant to tetracycline. Similarly, the bacterium causing the venereal disease gonorrhoea produced, as a result of selection, strains resistant to sulphonamides; more recently strains with partial resistance to penicillin and streptomycin have emerged. Figure 13 shows the dramatic rise in infections by resistant strains of *Salmonella typhimurimum*, the causative organism of typhoid.

bacterial resistance to sulphonamides

We can see with the benefit of hindsight that the incidence of this resistance might have been much reduced, or delayed, by not prescribing antibiotics unless really necessary and, more important, using larger doses in the first place so that the chance of there being survivors was reduced. The chances of the double mutation (to give resistance to large doses) having occurred spontaneously without the effects of antibiotic selection are very small. If the first mutation carries no advantage, and remains rare, the chances of the second mutation occurring among those few individuals is very slight. Thus, if we had used, say, 100 rather than 20 mg/litre of streptomycin in our cultures of *E. coli*, we would not have expected any survivors in the Petri dishes.

In modern medicine, therefore, it has to be accepted that any pathogenic bacterium may give rise to strains resistant to almost any antibiotic. This means that the situation is always a dynamic one. New strains evolve; new antibiotics are produced to beat them. Slight changes in the structure of the drug are sometimes sufficient, and by an empirical process of 'molecular roulette' and testing, research departments in the pharmaceutical industry and elsewhere have been able to keep ahead of the adaptive changes in the bacteria. This has been discussed in the radio programme of Unit 10, so you may recall how this has been achieved at the molecular level, particularly by the production of new penicillins.

## DDT resistance

A similar selective process has occurred among several insect species. As early as 1947, reports began to appear of strains of the housefly resistant to the chemical dichloro-diphenyl-trichlorethane (DDT). This chemical is poisonous to most insects in very low concentrations, and its introduction as an insecticide two or three years earlier had marked a real breakthrough in pest control. However, in various areas in countries all over the world, populations of flies soon appeared which were not killed by DDT in the concentrations originally used. As with the bacterial example given above, the dose had to be raised to be effective, but this was only possible within limits, as very large doses of DDT are not safe for either man or domestic animals (indeed, at the present time, it is not considered desirable even in quite small doses). As a result, other chemicals have had to be developed to replace DDT in many parts of the world.

**DDT resistance in insects**

Under laboratory conditions, a significant degree of resistance has been evolved within three generations in the housefly, by a process similar in principle to that used to produce antibiotic-resistant bacteria.

DDT-resistant strains have also appeared in many other insects, including some mosquitoes, body lice, bedbugs and cockroaches. Some insects have even evolved strains resistant to poisons such as prussic acid and lead arsenate.

## Resistance in mammals

It is not only such prolific organisms as bacteria and insects that are able to adapt fast enough to embarrass us in our attempts to exterminate them. Recent instances involving mammalian pests provide similar examples.

For some years the Brown Rat (*Rattus norvegicus*) has been very effectively controlled by a poison known as Warfarin. This contains a substance called di-coumarol, which, if it is eaten regularly, interferes with the clotting processes of the blood. If this happens the rats become, in effect, haemophiliacs†. They gradually weaken, and finally die from minor bites or bruises, or when having litters. They do not associate their condition (and that of those around them) with the food, and continue to eat it, which is why the poison is so useful. (Where acute poisons such as phos-

phorous are used, the survivors become wary, and stop eating the bait and so cannot be destroyed in the same way.)

In 1960, strains of rats appeared in Shropshire and Worcestershire which were unaffected by di-coumarol and indeed were flourishing on the poisoned bait. The incidence of the resistant individuals in the populations of these counties and of parts of Wales had risen to about 50 per cent in 1970, presumably because of the intense selection pressure provided by the widespread use of Warfarin. An interesting aspect of this resistance is that it appears to be due to a single mutant gene, and individuals which are homozygous for this gene are in some other respects weaker than normal (Warfarin-sensitive) rats. We will return to this aspect in section 19.4.4.

Warfarin resistance in rats

The physiological details of this resistance are poorly known, but, in spite of intensive efforts by the authorities, this strain (or strains) is now spreading fast towards the industrial Midlands. Attempts to contain the spread by surrounding the areas with a belt in which old fashioned methods of extermination are intensively applied have not been successful. If the new strains reach the large towns, the rat population, already more numerous than the human one, is likely to rise dramatically.

A second example is provided by the rabbit. This animal has been something of a pest of arable crops and grassland in Europe (Unit 20) for two hundred years, though offsetting its nuisance value somewhat by providing a source of food and felt hats. It became a far more serious pest after its introduction into Australia, where the tremendous efficiency of its reproductive system compared with that of the native pouched mammals (Unit 21) enabled it to multiply almost unchecked.

Attempts were made both in Australia and Europe to control it by the introduction of a virulent virus disease, myxomatosis, which is endemic among a species of rabbits in South America. This was very nearly successful; in many areas well over 90 per cent of the rabbit population was destroyed. It looked for a time as if the level would fall so low that the animal would indeed become extinct over wide areas. However, in Britain at least, it became apparent that the disease was not hitting all the population, because a small percentage of it was, for some reason, living above ground in nests, rather as hares do. Rabbits of course normally live in large crowded warrens underground and, as the transmitting agent of the disease in Britain is the rabbit flea, the disease spread fast. It became, however, a very effective selective agent favouring those individuals who lived relatively solitary lives above ground—they only rarely caught each other's fleas, and thus each other's myxomatosis. (Under normal conditions they were probably less successful, this open-air habit leaving them more at risk from predators.)

myxomatosis

This resulted in something of a rally in the rabbit population by individuals who were not necessarily physiologically resistant, but whose behaviour protected them to a considerable degree.

In the Australian rabbit population however, it appears that a genetically resistant strain emerged after the initial epidemics. The investigation described below (Table 4) was undertaken on rabbits from the Lake Orana district. Precautions were taken to ensure that none of the rabbits tested could have acquired immunity by having caught and survived the disease (in the same way as you may have an immunity to measles as a result of having had it in childhood), or by receiving a temporary 'inoculation' against it from their mother's bloodstream before they were born.

resistance to myxomatosis

In order to make sure that the change was in the rabbits and not in the virus (see below), the virus used was from the original outbreak. It was stored, and each year some of it was tested on laboratory rabbits to confirm that its virulence had not declined.

**Table 4**   Investigation showing the emergence of a strain of rabbits resistant to myxomatosis

| Test rabbits, taken from the wild population | Numbers of epidemics of myxomatosis previously suffered by the population | Symptoms | | |
| --- | --- | --- | --- | --- |
| | | Fatal % | Moderate % | Mild % |
| 1  Prior to arrival of myxomatosis | 0 | 93 | 5 | 2 |
| 2  in 1953 | 2 | 95 | 5 | 0 |
| 3  in 1954 | 3 | 93 | 5 | 2 |
| 4  in 1955 | 4 | 61 | 26 | 13 |
| 5  in 1956 | 5 | 75 | 14 | 11 |
| 6  in 1958 | 7 | 54 | 16 | 30 |

In Britain, although the mortality declined in the later epidemics, there seems to be no evidence that this was due to a genetic change in the rabbits. Apparently a mutation occurred in the virus which made it much less often lethal to the rabbit. This quite frequently happens with disease-producing organisms, and because such a mutation is advantageous to the organism the mutant form is likely to become the predominant strain. (Clearly a parasite which kills its host too soon commits suicide, whereas if a balance can be struck it can remain in the population indefinitely.) If its victims survive an attack they will acquire a degree of immunity to it; and if this happens to many individuals, even an epidemic of the original strain will probably not kill them, as they will have at least partial immunity to this as well.

*mutation of mxyomatosis virus*

**Would the immunity acquired by a survivor be passed on to its young?**

No, it is a phenotypic change only.

(Temporary immunity may be passed on by a mother to her young from her blood system, before they are born, or through her milk; but this immunity is not only transient, it is really only a temporary 'inoculation' by the mother—nothing to do with the bodily capabilities of the young themselves.)

These examples illustrate clearly two general points, the first being the fundamental one, that selection (in these cases as applied by man) will act on variation to produce a suitably adapted population. The second follows by implication: this adaptation means it is foolish to expect a 'wonder drug' to solve a pest or disease problem indefinitely. The situation is a dynamic one and unless this is appreciated from the outset, only temporary relief will be obtained.

### 19.6.2   Adaptation to exploit new food sources

If new food sources are introduced into an environment, any organism able to exploit them will be at an advantage. Sometimes this ability requires only a change in behaviour, but in other cases profound physiological changes are needed.

Many man-made substances, for example plastics, are potentially rich sources of energy; but they are new molecules so far as living systems are concerned, and so would require new and specialized enzymes to digest and metabolize them. They are also often long, condensed molecules, and these are anyway difficult to attack—few organisms can digest either hair (keratin) or cellulose, both of which have been around for a long time.

It is probably true that the more complex an organism's nutritional system has become, the less likely it is to be able to evolve completely new

digestive and metabolic pathways. This, coupled with the very rapid generation time of many simpler organisms, means that we would expect to see any evidence of the ability to exploit these molecules arise first of all among so called 'lower' organisms. In fact, this appears to be the case. Whereas both rats and human babies have chewed plastics since they have been in general use, so far only simple moulds seem to have been able to use them as food. Poly-vinyl-chloride (PVC) has been considered absolutely immune to rot of any kind; but recently thousands of pounds worth of damage has been done to the PVC insulation of stored army electronic equipment by a mutant mould. Another surprise for the Armed Services came when the shock-absorbent undercarriage of some naval aircraft suddenly went rigid. This proved to be caused by millions of micro-organisms clogging valves in the hydraulic system—they had been feeding on the hydraulic fluid, which is made up basically of hydro-carbons.

There are now mutant moulds which will attack the plastic in household emulsion paints, so that fungicides must be added to prevent your bath-room walls becoming a culture medium.

The examples are many, but the message is the same. Whenever a mutation gives rise to an enzyme which will allow a new molecule to be used as a food source, the advantage that this confers on the possessor may well lead to its multiplication throughout the environment. This can be checked by the addition of poisons, but there is always the likelihood that resistance to these will arise. Again we can see that the situation is a dynamic one.

## 6.3 Deliberate selection

Possibly the most important examples of the selection of variants, as far as human beings are concerned, are found in the field of plant and animal breeding. Nowhere near the present world population of human beings could be fed had not 'artificial' (i.e. human) selection produced new species and strains of crops and domestic animals. For example, the yields that could be obtained from wild cereals, if these were planted in the acreage today available to man, would satisfy less than half his present energy requirements.

The prognosis of the food supply and demand situation is somewhat depressing; even so, it leans heavily on the assumption that selective breeding will continue to increase productivity at least as rapidly as it has done over the last fifteen years or so. Thus, although the most familiar examples of selective breeding may be found among dogs, or in Shetland ponies, shire-horses and racehorses, the most important examples are to be found in such instances as the transformation of wild ancestral maize and rice into the modern high protein varieties, and the production of wheats resistant to attack by rust fungus.

**selection of crops and domestic animals**

This is not to say that the selective breeding of animals is without impor-tance, particularly to those of us in countries where animal protein forms a large part of our diet.

A recent survey comparing poultry production in the 1960s to that in the 1930s showed that a modern bird not only grew much more rapidly, but that it required only half the food per pound gain in weight compared to the 1930 bird. A new breed of pig has just been produced which reaches bacon weight in 150 days instead of the current norm of 185 days; and on 20 per cent less food. The genetics of this sort of change are immensely complicated, as many different genes are involved, but the approach is much the same as in the development of new crops.

Until such time as genetic material can actually be manipulated to give man what he needs, the only available method is the careful and informed

selection of mutants or recombinants. The crude mutation rate can be accelerated, but this is seldom helpful in practice. However, if the relevant genes can be located on the chromosomes, the process of recombination can be shortened. Many of the techniques employed are very sophisticated, but the principles involved are those you are already familiar with.

### 19.6.4 Apparently anomalous selection

In order to predict the likely outcome of a particular selective pressure on a population, it is necessary to have a fairly complete knowledge of all the factors involved. This information will be very difficult to obtain for a natural population in the wild. Consequently, the adaptive response observed may appear distinctly unexpected.

Just such a situation has been observed during studies of the disease sickle-cell anaemia, described in section 19.2.1. You may recall that this was caused by a mutation which resulted in the incorporation of an incorrect amino acid at one point in the protein chains of the haemoglobin molecule. You may also remember that the mutated gene known as Haemoglobin S (or Hb.S) is recessive; thus only those persons having the sickling gene from both parents (i.e. homozygous for the gene) actually suffer from the disease. These sufferers, however, are gravely disadvantaged, and have only a 20 per cent chance of surviving to maturity as compared with normal babies in their community (Fig. 14).

**sickle-cell anaemia**

> **What would you expect to be the fate of a mutated Hb.S gene in a population? Would you anticipate that it would:**
>
> **(a) spread steadily throughout the population as time went by;**
>
> **(b) rise to a particular frequency in the gene pool of the population say 20 per cent and then level off and remain constant;**
>
> **(c) be slowly eliminated as it appears in the phenotype (i.e. when inherited from both parents)?**

You would expect it to be eliminated as it became frequent enough to start appearing in both of the homologous chromosomes. The disease is a very harmful one, so that the gene is almost a lethal mutant when homozygous. However, *see the following text.*

It is clear that we would not expect the mutation to be favoured; indeed a harmful mutation of this sort is normally eliminated. This will happen rapidly where it is a dominant one, more slowly where it is recessive, as selection can only act on a trait when it appears in the phenotype. New mutant Hb.S genes would obviously arise spontaneously, so they would always be present in the population, but a substantial percentage of them should be eliminated at each generation. However, investigation does not bear out this expectation. Confining our attention to Africa and Asia, where most of the work has been done (the disease also occurs in some Mediterranean areas and the West Indies), we can see that there are large areas where the gene occurs (usually in a single dose, of course) in 15–20 per cent of the population (Fig. 15). Within these areas there are even communities with averages up to 40 per cent.

The reason it is possible to measure the frequency of the gene in the population (thus getting the information given in Figure 15), although the character is recessive, is because, even when the chromosomes are heterozygous for the mutant (i.e. inherited from only one parent), its effects can under some circumstances be detected. The bearer does not show the disease, but under conditions of low oxygen pressure, produced by very high altitudes, or artificially in the laboratory, the blood cells will show some sickling.

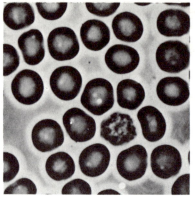 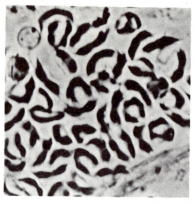

*Figure* 14      (*a*) *Hb.S.Hb.S*                              (*b*) *Hb.S.Hb.S*
                           *High $O_2$ level*                                       *Low $O_2$ level*

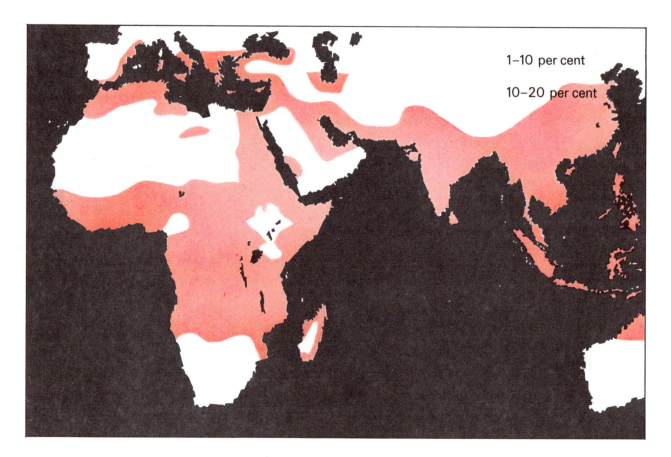

1–10 per cent

10–20 per cent

*Figure* 15   *Map showing the distribution and frequency of the sickle-cell gene.*

Normally, unless the bearer is a mountaineer, there is no visible effect on the blood of the phenotype.

Clearly the gene is not being removed from the population at the rate (16 per cent per generation) which should apply at this frequency. Not only does direct observation suggest that the levels are fairly constant, but the mutation would have to have occurred with quite incredible frequency sometime in the recent past to account for the levels observed on the basis of selective disadvantage. Thus, in spite of what we have said about the debilitating and lethal effects of the disease, we can only conclude that in some way the Hb.S gene confers a considerable positive selective advantage over the normal Hb gene, enough to offset the losses of genes at each premature death of a phenotype from anaemia.

43

**Can you think at what stage this advantage could possibly be felt? The clue lies in what we said in the section of genotype/phenotype relationships section 19.3 and above, when discussing the heterozygous condition.**

Thus, investigators concluded, as you may have done, that possession of a single Hb.S gene conferred some selective advantage that outweighed the harmful effects of the double dose. A possible answer may be found by examining a map showing the areas with a high death rate from malignant tertian malaria. Compare this with the known incidence of the Hb.S gene. Malaria is a disease caused by an organism which lives and multiplies in the red blood cells.

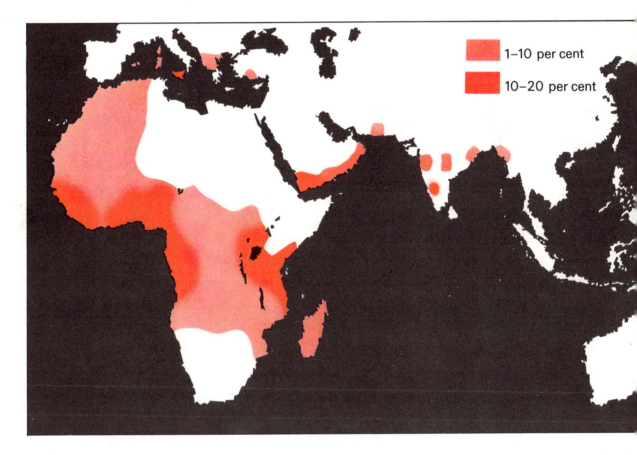

1–10 per cent

10–20 per cent

*Figure* 16   *Distribution of malignant malaria.*

This suggested to the investigators that a single Hb.S gene might provide resistance to the malarial parasite, perhaps at the stage when it is normally living inside the red blood cells and destroying the haemoglobin. Their research has largely confirmed this. In these areas, malaria is almost always present and the child mortality from it is high. (If the children survive it they normally acquire considerable immunity.) Children with a single Hb.S gene were shown to have a 25 per cent better chance of surviving attacks of malaria than those with the normal genes. There is further confirmation. There are regions within the malarious areas where the conditions do not favour the mosquitoes which transmit the malaria. In these regions, malaria is not much of a problem and, as expected, the Hb.S gene is a rarity.

It emerges, therefore, that selection is not acting to remove the genes causing sickle-cell anaemia, as might have been expected. They are indeed removed when homozygous, but this selection is outweighed by their selective advantage when heterozygous, due to the effect of malignant malaria.

44

In summary:

(1)  Hb.Hb  
*normal phenotype*  
*susceptible to*  
*malaria*

(2)  Hb.Hb.S  
*normal phenotype*  
*resistant to*  
*malaria*

(3)  Hb.S.Hb.S  
*anaemic phenotype*  
*usually dies*

'Malarial selection' favours (2) but, as the frequency of the gene rises, (3) will appear more commonly, and die off. Therefore a balance is reached, in which the frequency of the gene will vary considerably with the intensity of the 'malarial selection' pressure.

If this interpretation of the situation in Africa and elsewhere is correct, our knowledge of population movements would suggest that the sickling gene has been in the population in this sort of balance for thousands of years.

---

**If you changed the environment of one of these populations with a 20 per cent frequency of Hb.S, in their gene pool, so that malaria was no longer a major cause of death, would you expect:**

(a) **that the frequency, now so well established would remain roughly constant;**

(b) **that the frequency would decline at each generation by a predictable amount, until it is eliminated altogether;**

(c) **that the frequency would decline at each generation by a predictable amount to a low level where the rate of elimination equalled the rate of production of new Hb.S mutants?**

---

(c) is what would be predicted.

---

The first stages of this have in fact been demonstrated. American Negro slaves were imported from West Africa in large numbers between 250–300 years ago. If the frequency of the Hb.S gene in the original population is assumed to have been not less than 22 per cent, and it is allowed that this figure would be reduced by interbreeding to around 15 per cent, it is possible to predict to what level the gene frequency should have fallen today.

In the absence of any selective advantage, after twelve generations the deaths from anaemia should have reduced the frequency to 9 per cent (which would result in 1 per cent actually showing the disease). This is, in fact, the frequency found among US negroes at the present time. On the other hand, West Indians in Britain (with a very recent change of environment) show an incidence of 18–20 per cent. This fact was not appreciated until several West Indian patients in the UK collapsed whilst having teeth removed under anaesthesia, because of sickling of the red blood cells. (The dental anaesthetics used may lower the $O_2$ level in the blood.)

*Warfarin resistance and selection*

It is interesting to speculate on whether a similar situation of 'balanced' selection pressures may not be operating in the case of the Warfarin-resistant rats mentioned in section 19.4.2 above. You may recall that the mutation which conferred resistance left them physiologically weaker than normal sensitive rats, when it appeared in the homozygous condition. This fact emerged when efforts were made to keep resistant rats in the laboratory to investigate the physiological basis of the resistance. The rats which were homozygous for this mutant gene died very easily and appeared to be deficient in vitamin K. It will be interesting to see if further work reveals that malaria resistance in man and Warfarin resistance in rats are similar illustrations of the same evolutionary situation: namely a mixed

45

population maintained by the balance of two potentially conflicting selection pressures.

### 19.6.5 The stabilizing effect of selection

An important point which has emerged by inference several times above is that natural selection does not necessarily act to produce change; it may well produce stability. Indeed, in the short term, it is far more often doing the latter, and this is of the greatest importance.

We have seen that, in fact, most mutations are likely to be deleterious to a greater or lesser extent. The genes thus produced will be eliminated by natural selection. The speed at which this happens will depend on whether the gene is dominant or recessive, and whether the new trait that the mutant gene gives rise to is merely slightly less suited to the environment than the original, or is downright lethal.

In a theoretical environment which remains constant, once a population has become well adapted, it would be reasonable to suppose that there will be no adaptive change, and that selection will be concerned *only* with maintaining stability by eliminating less fit variants. In practice it probably never comes to this. Environments are not constant. If the physical surroundings remain constant, there will still be changes in either the predators, the prey, the parasites or the competitors. Evolution appears to be a continuous process, so no environment is likely to be by-passed indefinitely. There are a few which have changed little, such as parts of the deep sea and some islands, and these may support such well-known 'living fossils' as the coelocanth, a fish which has scarcely changed since the Cretaceous, and the tuatara, a very primitive reptile with a bone structure which has remained unchanged for 170 million years (see also *Understanding the Earth*). The tuatara has now to be vigorously protected from competition with modern mammals. On the other hand, a few forms are so efficient that they seem to be able to survive considerable changes in the environment without having to undergo much more adaptation themselves—for instance, many insects appear much as they did in the Cretaceous period, 120 million years ago.

**'living fossils'**

So the pattern is usually one in which selection maintains the stability of a well-adapted population within a reasonably constant environment. Whenever environmental changes do occur, however, selection will lead to adaptive change in the population, provided always that variants are produced in sufficient numbers. If, however, *too many* variants are produced, the stabilizing effects of selection will be removing too many 'unfit' individuals, with the result that the population will decline. Natural selection has of course favoured the optimum balance, and herein lies the risk of raising the mutation rate in human populations by increasing the doses of radiation they receive. A population with too little variation may die as the environment changes; but one with too much may die in the existing one.

Much of this stabilizing selection will occur before the variant sees the light of day. For example, major genetic changes may result in the death of the zygote during the first few divisions—in the case of mammals before it has even implanted in the uterus. Structural changes of various kinds may put the embryo at risk at a later stage and spontaneous abortion may occur. It is estimated that twenty per cent of the human eggs successfully fertilized do not in fact survive until normal birth. This process is sometimes called internal selection; but it is the same process that we have been discussing all along. Fitness to obtain nourishment from the wall of the uterus is no different in principle from fitness to digest the mother's milk once out in the world.

## 9.7 Speciation

### 7.1 Species

We have seen how selection of spontaneously occurring variants may lead to changes in a population, sometimes quite major ones; but in none of the examples given did we claim that the changed population no longer belonged to the same species. Yet it is obviously central to the theory of evolution by natural selection that such processes *do* lead to the formation of completely new species, and it is quite logical to suppose that this indeed happens. We have not yet demonstrated it for two reasons, one of time-scale and one of definition. To merit the description of a new species, a population will have to undergo a whole series of changes, usually involving both structure and behaviour, which with most organisms will involve a much longer period of time than we have been considering. In addition, we are brought up against one of the most difficult points in any discussion of speciation, namely 'what constitutes a species?'

*The taxonomic view of species*

The science of taxonomy, the classification of organisms into groups with clearly defined similarities and differences, really owes its beginning to the work of Linnaeus (1707–1778). He ranked organisms according to the degree of anatomical similarity, with species as the basic fixed units. These corresponded to the popular 'kinds' of animals—a 'cat' an 'elephant' and so on—and he gave each a Latin name of two words. The first word was the *generic* name and the second is the *specific* name. Where species had **species** obvious affinities, for example cat and puma, he grouped them in the same genus. This is still the system, the cat being *Felis maniculata*, the puma, *Felis concolor*. Minor varieties or races might be denoted by adding a sub-specific name after the specific one. For example, the English wren is *Troglodytes troglodytes troglodytes*.

As our knowledge has grown, the number of the hierarchical ranks originally suggested by Linnaeus have to be increased; but the principle remains the same. A complete classification of a species will involve placing it within a system of twelve steps or ranks, denoting degrees of affinity. However, some of those ranks are met with very much more frequently than others. The smallest grouping of species (one which indicates those species with a large number of characters in common) is into a genus, as indicated above with reference to some of the cats.

Another example is modern man, whose specific name is *sapiens*. He belongs to the genus *Homo*, which he shares with a number of fossil ancestors (mentioned again in section 19.8). This genus shows close affinities only with other extinct genera, and it is grouped with these into the family *Hominidae*. This family is grouped in turn with two others, which contain the gibbons, chimpanzees, orang-utan and gorillas, into a superfamily, the *Hominoidea*. This process is continued until man's ranking with all other living organisms is described.

This system of classification has been extended to embrace over 1.5 million species, distinguished primarily on anatomical features. These include:

| | | |
|---|---|---|
| *Vertebrates* | Mammals | about 4 000 species |
| | Birds | about 9 000 species |
| | Reptiles and Amphibians | about 6 000 species |
| | Fish | more than 20 000 species |
| *Invertebrates* | | 1 050 000 (including 850 000 insects) |
| *Plants* | | about 400 000 |
| *Unicellular organisms* | | more than 75 000 |

It may be deduced from these numbers that the process of describing and determining these species is one for the specialist. Indeed, such a detailed knowledge of the anatomy and life-history of an organism and its relatives is required for this purpose, that the field of expertise of a taxonomist may be a very narrow one.

This approach to the definition of species is one of several, and possibly it is a rather more static one than that adopted elsewhere in this Unit. Nevertheless it is a view of great practical importance. In the practice of both pure and applied science, it is necessary to be able to refer to a group of organisms in such a way that everyone concerned is clear as to the precise limits of this group. For example, it may make a great deal of difference to know which of two rather similar insect pests is attacking a crop. Differences in some stage of their life cycle could call for completely different control measures.

Furthermore, the taxonomist's species do really exist. These species may live side by side and yet remain distinct, as, for example, do 200 species of fish of the genus *Haplochromis* in Lake Malawi (Nyasa). As you will see in Units 20 and 21, such groups live and interact with one another in predictable ways, the interacting units being discrete species.

Thus the work of Linnaeus began to bring order, precision and system into descriptive biology; but his system was to some extent arbitrary, and misleadingly simple. This was because it was based simply on structure, and it also assumed that species were immutable. The Creator had made all organisms at the beginning and they stayed that way. Thus the similarities between the cat and the puma were due to a whim, albeit an Almighty one. (Presumably on that day all the carnivorous animals were being run off to a similar pattern.) Species were obviously the fundamental unit because they were clearly distinguishable and unchanging. Today a classification has to represent something more than anatomical similarity; it must indicate common origins and family relationships. Nevertheless, Linnaeus' judgement was such that most of his species still stand; even though they may be grouped into different genera.

**From what you know of evolutionary mechanisms would you expect to be able to classify all organisms into distinct species in Linnaean terms?**

No; once you begin to collect organisms from all over the world, you find some intermediate populations having some of the properties of one species, but with others usually ascribed to a related species.

Both Lamarck and Darwin, who assembled immense collections from regions all over the world, often found intermediates they could not reasonably say clearly belonged, on anatomical grounds, to either one or other of two related species. This confirmed their belief in the view that species evolved from other species. The problem of definition is therefore acute. An intermediate can be described as a 'race' or sub-species of one

or other of the species, but at what point must we elevate this race to a species? If biological evolution is the dynamic process we think it is, then 'species' must be a dynamic concept and we must accept that a species exists only for a period in time. Obviously then, it is a term of convenience (even if an essential one for the rational study of biology), and the lines of demarcation will be to a large extent biologically arbitrary. What is more, what constitutes a satisfactory specific character may vary with the field of study of the biologist concerned. To an anatomist or museum-based taxonomist, physical characteristics such as size and details of anatomy may be the most important; to someone studying organisms in relation to their environment, it may be their habits; to a geneticist, it is the frequency of particular genes in the genotype.

There is, however, one common factor involving all these approaches to the problem of the distinction between race and species. This is whether or not the race or variant population interbreeds with the 'original' or 'main line' population. If it does, we would expect to find a continuous physical gradation from the 'norm' of the original population to that of the race. If they meet and mate under natural conditions, their habits and behaviour must also still be closely related. Lastly, and most important, if they are exchanging genes they are effectively 'diluting' the differences which selection may be increasing. Thus, if they are interbreeding to a significant degree, it is reasonable to consider that the variant population is still only a race of the original.

**failure to interbreed**

The other side of this coin is that the major step in species formation is when a race becomes reproductively isolated from the parent population.

### 19.7.2  Gene pools

In section 19.4, we described the evolutionary unit as the 'freely inter-breeding population', the unit on which selection can be seen to act. We can adopt a narrow genetical view of this unit, disregard the phenotype altogether and consider only the genotypes. In fact, it is convenient to go still further, and regard the population only in terms of the gametes it produces (the haploid germ cells), a few of which will fuse with one another to form the new generation. This rather theoretical entity is called a *gene pool*; in the terms in which we defined a population, every gamete (or gene combination) in a gene pool should have an equal chance of fertilizing any other.

**gene pool**

In 1909, a geneticist, Hardy, and a mathematician, Weinberg, produced a simple equation which predicts the distribution of genes within such a gene pool, under ideal conditions. The Hardy-Weinberg equation forms the basis of almost all modern population genetics. It states, in effect, that for any pair of genes controlling a character, one dominant and one recessive, their relative frequencies in the gene pool, whatever they may be, will, in the absence of factors which modify gene frequency, remain constant from generation to generation. The assumption is made that the population is sufficiently large for sampling errors to be ignored.

The equation gives a precise answer to the problem exercising the early evolutionists, including Darwin: namely, why sexual reproduction does not lead to increasing uniformity. The proportions of variants are maintained with a random distribution—some individuals homozygous for one gene, a majority heterozygous, and some homozygous for the other.

The main modifying factor is selection, and where a gene is favoured or discriminated against, the equilibrium will be shifted. In addition, if a gene has a high mutation rate, or if significant numbers of phenotypes carrying it migrate in or out of the population, the equilibrium will be disturbed. However, the Hardy-Weinberg prediction will only apply where there is

reasonable intermixing. As soon as part of the population ceases to exchange genes freely with the rest, there is a probability that divergence will take place.

### Reproductive isolation within gene pools

This can come about through various circumstances, and the effects are progressive. Gene flow from one side of a large population to the other may be interrupted by geographic factors—i.e. 'barriers' such as mountains, deserts or water—or simply by the discontinuity of the population in an area. If disease, or fire, or a predator effectively wipes out the population in an intermediate area, recolonization may take time. During this time, the 'outlying' population will exchange few genes with the rest (Fig. 17). If conditions differ substantially, the effects of selection on the relative frequencies of the genes in the two groups may become apparent.

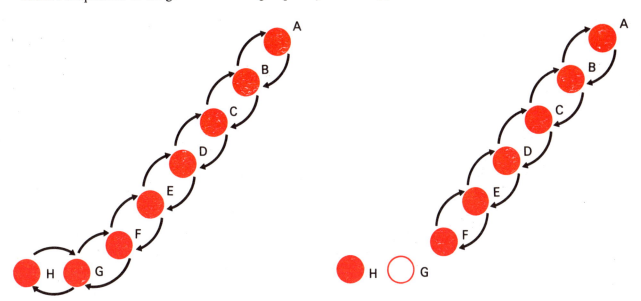

Figure 17 *Interruption of gene flow across an interbreeding population.*

Any of these factors will obviously encourage race formation, or, where it is already present, slow down exchange between race and parent population. The crucial factor, however, is that once the differences are there, reproductive isolation will tend to increase, thus further increasing divergence.

An actual example of this is provided by the distribution of the breeding ranges of the thirty-four 'races' of the American Song Sparrow *Melospiza melodia* (Fig. 18). Where the bird occurs in mountainous regions or on offshore islands, distinct races have become isolated from one another, each sometimes having only a few square miles of territory. Where these geographical barriers are less pronounced, a single race may have a territory of thousands of square miles, across which gene flow can occur relatively unimpeded.

In Unit 21 you will find an example (Darwin's finches) where this process has proceeded a stage further.

### Isolation maintained by behaviour

Small differences in the markings of birds will reduce their attractiveness as mates to members of the original population. (See the offprint with this Unit.) Changes in song patterns may have the same effect. If an animal changes its feeding habits it meets and mates less frequently with the main population. (Races of mosquitoes apparently sharing the same environment may never meet for practical purposes if one group develops day

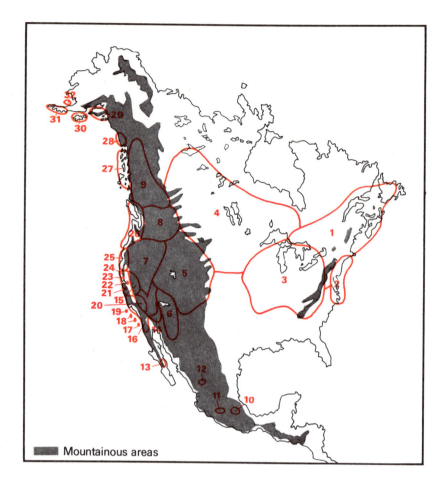

*Figure* 18    *Distribution of races of* Melospiza melodia *in North America, showing main geographical features.*

flying and feeding habits and the other evening or night flying ones.) These, and many other small differences, may effectively isolate one race or part of a gene pool from the rest, and selection will increase the divergence. Thus they are no longer part of the same pool. But are they a new species?

The 'touchstone' usually employed in deciding whether a variant population constitutes a race or a separate species, is to see if they will interbreed with the 'parent population' and produce fertile offspring. Where the genetic differences have become marked, the offspring (called hybrids) are normally infertile. This is due to various factors, often the inability of maternally and paternally derived homologous chromosomes to 'recognize' one another and pair up correctly at meiosis. Many such cases of hybrid sterility are well known. The mule (horse × donkey) is a case in point. It must be bred from horses and donkeys each time, not from other mules. In this case, reproductive isolation is complete; no genes will be exchanged between horses and donkeys. Sometimes it is only a question of degree. Domestic cattle can be crossed with American buffalo, yaks and other relations. The hybrid bulls are sterile, but the cows are fertile, and will backcross with domestic bulls, buffalo or yak. In the case of two particular species of *Drosophila*, there is merely a lowered fertility in the hybrids; but this, coupled with other isolating factors (they have rather different habitats within the same area; the males prefer the females of their own species; and they are most active at different times of day), means that under natural conditions they do not exchange genes. Thus, in this case at least, four partially isolating factors operate cumulatively and lead to complete natural isolation. So even failure to interbreed is not a simple and absolute criterion to use to distinguish separate species.

**hybrids**

51

Although the two species of *Drosophila* referred to above can in fact still exchange genes (though they seldom do), they are considered as discrete species. Similarly the races of *Melodia* can interbreed, but usually do not, yet they are considered to be a single species. If you, yourself, could sit on a California hillside for a hundred years, watching these birds, you would probably observe only minor changes in the race. On your return to civilization, however, it is possible that you would find that you had witnessed the evolution of a new species.

From what you have read in this section, then, you will not be surprised that it has not been possible to say that 'species A1 has been shown evolving from species A by a process of natural selection'. There is much evidence from which we can infer the truth of such a statement, for example, the evidence from micro-evolution, and the observable changes in gene-frequency in populations. But our own time scale, and that of the development of the theory itself, makes it unlikely that we could ever claim to have actually witnessed the event.

We suggest that at this point you should read the set text, which for this Unit is an offprint from the journal *Scientific American* of October 1967. It is an article by Dr. N. G. Smith describing work he has done on the reproductive isolation of four closely related species of gulls, nesting in overlapping territories. In addition to providing an illustration of one of the points made above, it is interesting as an example of the type of experimental situation with which a biologist may be confronted. You may feel that travelling between Arctic cliffs in a canoe, carrying tranquillizers and a paint box has nothing in common with manipulating electronic devices in the laboratory. Yet it has; the logical processes, the formulation and testing of hypothesis, analysis of the data and checking the significance of the figures, are all common to both. The main differences probably lie more in the temperaments of the investigators.

We have offered you a few simple questions which you may find useful. In two of them we ask you to comment critically on some of the statements in the article. (An attitude we hope you will adopt to all our material.) If you feel that *you* would have written them differently, bear in mind that this is an article from a magazine with a mixed readership, not the publication of detailed results in a specialized journal.

**What was the basis of Smith's claim that pair formation depended on female initiative?**

See Answer 9, p. 77.

**Which of the following factors does this work show to be *irrelevant* to successful copulation in established pairs of gulls?**

(a) **Soliciting behaviour of the female.**

(b) **Wing-tip pattern of the female.**

(c) **Eye contrasts of the female.**

(d) **Wing-tip pattern of the male.**

(e) **Eye contrasts of the male.**

(f) **State of development of the male's testes.**

See Answer 10, p. 77.

**On p. 7, column 3, paragraph 1, Smith says his original working hypothesis on why pairs stayed together before egg laying was that 'the main component of the pair bond was the attachment of the individuals to one another'. Would you consider this a useful hypothesis?**

See Answer 11, p. 78.

On p. 7, column 1, Smith refers to natural selection in the following terms. 'To avoid mixed pairings, then, selection favoured dark-eyed individuals where Kumlien's gulls, etc. . . .'

(a) Do you consider it legitimate to think of natural selection in these terms?

(b) It is suggested on p. 101 and p. 102 (bottom of column 1 and top of column 2) that natural selection is maintaining differences in the populations which tend to discourage interbreeding and the formation of hybrids. What underlying (but unstated) assumption must Smith be making about the hybrids, if his claim is true?

See Answer 12, p. 78.

Do you feel that this work has established a major cause of reproductive isolation between the four species concerned in terms that should satisfy a scientist? If not, what further questions would you suggest Smith should have tried to answer?

See Answer 13, p. 78.

## 19.8  Human Evolution

This is also the topic of the radio programme of this Unit. Human evolution provides an interesting and slightly unusual example of selection at work. If you can remember as far back as Unit 1, you will recall we suggested that man was remarkable for the development of his brain, and that a significant increase in brain size had occurred since the time of his ancestor of half a million years ago, *Homo erectus*. It is striking that in man's general bodily form there has been almost no sign of specialization over more than three million years, other than changes in the pelvis and foot allowing upright walking. In fact, man is built on all-purpose lines. He can run, crawl, dig, climb and swim; he has excellent vision (day and night), reasonable hearing and a fair sense of smell. (No other animal can do all of these things as well as man, though each of them can be done very much better by one specialist or another.) But while the body has not increased greatly in size since *Australopithecus*, a man-ape of a million years ago (whose average height was about four foot, six inches), the cranial capacity has risen by 300 per cent. (See Table 5, Fig. 19, p. 55, and the Radio Broadcast Notes.)

*increase in skull capacity*

**Table 5   Cranial capacity in human evolution**

|  | *Capacity (in cm$^3$)* |
|---|---|
| Australopithecines ('man-apes') | 450–550 |
| *Homo erectus erectus* (Java man) | 770–1 000 |
| *Homo erectus pekinensis* (Peking man) | 900–1 200 |
| *Homo sapiens neanderthalensis* (Neanderthal man) | 1 300–1 425 |
| *Homo sapiens sapiens* (recent) | 1 200–1 500 |

What then is the selective advantage which has resulted in this change? Clearly having a larger head has no advantage *per se*; it merely requires larger hats.

No simple direct relationship between skull capacity and mental capacity has been shown among living humans, and of course we cannot prove that modern man has greater intellectual potential than Java man. However, there is evidence that, between species, more brain cells in the cerebral cortex make for a more extensive memory, and also, as you will remember from Unit 1, a proportionate increase in the brain/body ratio leaves more nerve cells free from the 'autonomic' control of the body (see Unit 18) and available for 'higher' functions. It is also known that memory systems seem to involve a very large number of cells and that relatively enormous areas of the brain are involved in controlling the fine manipulation of the hands, tongue and lips.

It is reasonable to assume, therefore, that the expression of this increase was the ability to use, and later make, tools and to develop sophisticated speech. The advantages of both attainments in controlling the immediate environment are immense and obvious.

Tool manufacture was a characteristic of the man-apes who inhabited Olduvai Gorge (Tanzania) a million years ago. The ability to make a tool for the job is a great advance on the habit of using something that happens

54

to be at hand, as do some apes and birds. The skilled use of tools has come with the development of a thumb which can be used in opposition to the fingers, a necessity for fine manipulation.

Speech confers great advantages over any simple communication system to a hunter in the planning and co-ordination of food catching, but again it calls for huge numbers of extra nerve cells. Thus it is reasonably safe to say that man's evolution has centred round the evolution of his brain, selection favouring the skills made possible by an enlarged mental capacity.

All the fossil specimens of *Homo* that are thirty thousand years old or less (e.g. Cromagnon man, Fig. 19) are 'modern' in all respects, that is to say they are not structurally distinguishable as a sub-species from ourselves.

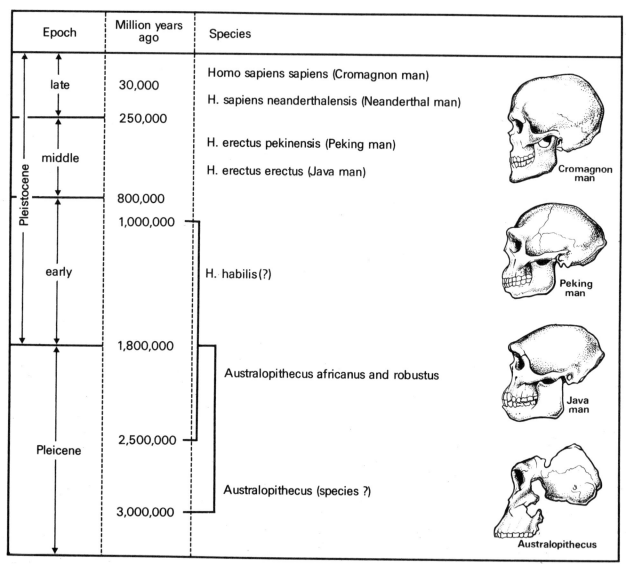

*Figure* 19

There is much evidence that their culture involved complex rituals and elaborate art, as indeed did that of some earlier groups; yet their basic 'economics' had apparently remained almost unchanged for several *million* years. Their stone implements were similar in function to those of *H. habilis* and the species of *Australopithecus* resident in Olduvai Gorge between two-and-a-half million and one million years ago. Like these latter species, *H. sapiens* was still a hunter and vegetable gatherer. (It is estimated that hunting accounted for about 25 per cent of his food; plant collecting about 75 per cent.) This sort of life requires a basically nomadic habit, and the land will only support a relatively low density of population.

The first evidence of any fundamental change in the way of life of *Homo* comes from what is sometimes called the 'New Stone Age', and, in particular, in the Middle-Eastern populations of about 8–10 thousand years ago. They show, for the first time, evidence of the domestication of animals and the cultivation of crops. This change is an essential one for the formation of larger human communities with a basically 'fixed abode'. It can be argued that human development remained merely an extension of this way of life until the Industrial Revolution. It is interesting to think that we may be witnessing the start of the third major change, the 'space age', to have taken place since the rise of *Australopithecus* between two and four million years ago.

### 19.8.1 Is man still evolving?

As the history of species goes, *Homo sapiens* is young; and it would be surprising to find that he had stabilized his gene pool by eliminating all the modifying factors discussed in earlier sections. Indeed, what we know of human genetics suggests that selection may be acting quite fiercely to change some genes.

**adaptation to modern conditions**

---

**Can you think of an example which we have mentioned?**

Sickle-cell anaemia.

The Hb.S/Hb combination is not favoured in the USA, or anywhere else where malaria is being controlled, and it is being eliminated from such areas.

---

The fact that we are living in an increasingly controlled and artificial environment merely means we must adapt to it, as we have had to adapt to changed conditions in the past. For example, genotypes sensitive to drugs such as penicillin and barbiturates (and they are not uncommon) will be eliminated as unfit, where previously they were neutral. The laboratory rat has changed so that it is unfit for life in the wild; but it is not decadent, merely adapted, as it is far better suited to life in the laboratory than is the wild rat. We may be moving towards a similarly enclosed existence, and selection may adapt us to it.

Of course, the situation is not always as clear-cut, since adaptation can only take place through reproductive fitness. For example, a 'virtue' in modern living may be the ability to resist new kinds of pressures, those caused by overcrowding, noise, pollutants and the more sophisticated social stresses. If, however, people have completed their reproductive lives *before* retiring into a mental hospital or dying of bronchitis, then selection cannot act. At the moment we really do not have the information to say whether those particularly susceptible to such things are reproductively as fit as the rest of the population or not.

### 19.8.2 Selection and intelligence

A most interesting example is the effect of selection on intelligence. Even if man's present dominance of his environment stems from his increased mental capacities, it does not follow that what is now needed for the benefit of mankind is greater intellectual advance, though many would assume that it is desirable. Nevertheless, the question remains: 'is selection favouring intelligence now, as we assume it did in our past?'

The first pitfall in assessing this is that we have no satisfactory definition of intelligence—it depends very much on the context in which it is being measured. Thus it can equally be said to be manifest in acts of character judgement, prudent cowardice or computer design. We commonly tend to measure the so-called Intelligence Quotient of a person by comparing his score in an intelligence test with a supposed national average performance of 100. The problems associated with this method are numerous, not the least being that what an intelligence test measures is the ability

to do intelligence tests, and many feel that there is only a marginal correlation between this ability and intellectual attainment. This has been nicely illustrated with rats. A group of highly 'intelligent' rats were produced by rigorous selection over several generations for their ability to run a maze. Thus, within four or five generations, they showed a marked shortening of the time taken to learn the maze, compared to that taken by controls or their own great-grandparents. However, it emerged that these faster-learning, or 'more intelligent' rats were only quicker at learning mazes. They showed no significant improvement in other types of learning and problem-solving situations. Thus 'intelligence' is not really an acceptable entity for study; it is necessary to specify a particular skill. But, in the absence of a better criterion, we still tend to use the skill of 'intelligence test passing' as our yardstick.

The second pitfall lies in a situation which is most obvious in the human organism: the interaction between the genetically-determined potential and the effects of education or training. We saw something of this when considering the athlete in the section on genotype and phenotype (p. 24). The potential mental capacity may be expected to be determined genetically, as with all other physical traits. However, the early environment seems to play a major role in deciding to what extent this potential will be realized. If this environment does not provide an adequate pattern of stimulation, development may be permanently retarded, even to the point where speech is difficult. Thus, although intelligence tests are designed to avoid testing formal learning, they cannot avoid measuring many environmental influences.

Having given these warnings, what of the results? There is a *prima facie* case for thinking that intelligence should be on the decline, at least in the Western world. It has been widely recognized, since the end of the seventeenth century, that throughout the West, and possibly China, the more successful members of the community have had fewer children than the norm. This is true whether 'success' is judged by social class, income or education. Not only would these people be expected to have higher than average intelligence quotients, but they would be expected to provide above average environments for their children. Measurements of I.Q. support this expectation. Thus, in terms of reproductive fitness, those with higher I.Q.s are at a disadvantage. This has been estimated to produce an anticipated drop of 1–4 points on the I.Q. scale per generation in the average level of the population; a fearful prospect.

There is very little evidence to decide if this expectation is being fulfilled. Only two really widespread comparisons over a sufficiently long time interval have been made, one on 88 per cent of all Scottish school children, with a fifteen-year interval (1932 and 1947) between measurements, and one on American soldiers in the First and Second World Wars, with an interval of some twenty-five years. The results of these showed that the scores on the more recent tests were, if anything, slightly higher; an unexpected and apparently anomalous result.

**Assuming the information we have given you as to the fertility of those with higher I.Q.s is correct, can you account for this result?**

One possible reason is education. We said that I.Q. tests measure education to some extent, and for the population as a whole this improved vastly in the periods under consideration. Another is that better diets can be shown to improve I.Q. and the national diet can be shown to have improved.

Thus it appears that if it is true that intelligence is being selected against, this is being more than compensated for, at the moment at least, by improvements in the general level of education, including perhaps at a pre-school level.

At the other end of the scale, another factor may be at work which could have a stabilizing effect on population I.Q. Recent work suggests that

marriage between low-grade mental defectives may produce only slightly more children than those between people of very high I.Q.

The highest fertility is to be found among the mediocre intelligence levels. So selection may, in fact, tend to eliminate the very stupid as well as the bright, thus maintaining something near the *status quo* (i.e. 'stabilizing selection').

**Table 6   Mean number of siblings of each inmate of a mental institution**

| Parents | Cases observed | Siblings (mean) |
|---|---|---|
| Superior × normal | 9 | 2.89 |
| Normal × normal | 798 | 4.72 |
| Normal × dull | 196 | 5.45 |
| Normal × feebleminded and dull × dull | 113 | 4.52 |
| Normal × imbecile and dull × feebleminded | 54 | 3.82 |
| Dull × imbecile and feebleminded × feebleminded | 24 | 3.58 |

Table 6 shows how many full brothers and sisters (siblings) there were of each inmate of an institution for mental defectives. The inmates have been grouped according to the I.Q.s of their parents.

All of this is sufficiently unproven to provide a fertile field of argument and clashing faiths. Current trends in human evolution are very difficult to distinguish, but evolution of some kind is undoubtedly taking place. What we *can* say is that today *Homo sapiens* is a single species, made up of several races all of whom share a large number of genes. These races are now exchanging more genes than they were even 500 years ago, when limited local colonization was probably the main cause. Today, almost every country has representatives of almost every other one living in it and to some extent intermarrying within it. Thus, in addition to cultural exchanges and changes, modern mobility is providing new genetic cocktails for new and existing selection pressures to work on. This will inevitably mean more change.

## 9.9  Summary of the Unit

1    To account for the great diversity of species, and the complexity of the adaptation they show to their environment, it is necessary to accept either that they were created as they are by a Divine Being, or that some process of organic evolution has taken place. The first interpretation is not examined in detail, but the second forms the substance of the Unit.

2    (a) If species have evolved, then it is clear there must be heritable variation as well as the continuity stressed in Unit 17. One of the ways the genotype of an organism may change is by mutation. This involves the re-arrangement or substitution of one or more nucleotides on the DNA of the chromosome. This change in the sequence of nucleotides results in a change in the polypeptide chains encoded in that part of the DNA. This will lead to changes in the proteins of the cell and thus visible phenotypic changes (e.g. phage resistance in *E. coli*, or its inability to manufacture tryptophan, defective haemoglobins in man, colour changes in *Sordaria* spores).

The main cause of naturally occurring mutations is not certain, but it is probably a chemical factor or factors within the cell. Many exogenous chemicals will raise the mutation rate, as will irradiation and, in some species, temperature rises. The timing of mutations appear to be quite random, but the rate varies between species and between genes within a species.

(b) Heritable variation also arises through sexual reproduction because of the fusion of gametes from different parents and because the haploid gametes are produced by meiosis. Meiosis involves the random assortment of the homologous pairs of chromosomes prior to division and, in many cases, crossing over. Thus the existing genetic material of a population is remixed and reshuffled at each generation.

3    Darwin laid emphasis on natural selection as the main factor leading to adaptation and evolution, with the inheritance of acquired characters and mutation as lesser factors. He postulated that overproduction within populations would lead to competition and that this would result in the selection of some variants more frequently than others, thus leading to a change in the population. Experiments can be designed to show whether selection acts on spontaneously occurring variation or on change actually induced by the environment.

4    Several examples of the micro-evolution that has taken place in recent years are considered. These include the selection of strains of pests and disease organisms which are resistant to the control measures used against them, and the deliberate selection by man of crops and stock showing certain characteristics.

5    Animals live in limited, interbreeding populations, and each genotype may be considered as contributing to the 'gene pool' of this population. The frequency of the genes in the gene pool will remain roughly constant so long as the members can interbreed, except where pressures such as selection act to change the frequency of a gene. If the members cease to interbreed freely, one part of the pool becoming isolated from

the next, then selection or other forces may cause divergence between the main pool and the isolate. Such isolation may become permanent if the phenotypic expression of the divergence even further discourages the exchange of genes.

This may lead to the formation of new species, or just distinctive races or sub-species. The borderline between species and race is difficult to define, but is usually drawn on the basis that definite anatomical differences together with a *de facto* failure to exchange genes will justify distinction at specific level. There are many exceptions to this generalization however.

6  Human evolution and the effects of selection on the human genotype are briefly considered.

Appendix 1 (White)

## Glossary

ACHONDROPLASTIC DWARFISM   A hereditary abnormality in which the limbs are much foreshortened, although the head and trunk remain normal.

ANTIBIOTIC   A substance, produced by a living organism, which diffuses into its surroundings where it is toxic to some other species. Closely related substances can now be synthesized and go under the same general name.

GENOTYPE   The genetic constitution of an organism (i.e. the particular set of pairs of genes in each cell of an organism). (See Unit 17.)

HAEMOPHILIAC   One suffering from the hereditary disease, haemophilia, in which the ability of the blood to clot is lost or impaired. In serious cases even minor internal damage can lead to fatal haemorrhage.

HETEROZYGOUS   Having two different genes (e.g. the normal gene for a character and its mutant) at the corresponding points on a pair of homologous chromosomes.

HOMOZYGOUS   Having identical genes at the corresponding points on a pair of homologous chromosomes.

PHENOTYPE   The sum of the characteristics manifested by an organism, as contrasted to the genotype. (See Unit 17.)

SPINDLE   A body which forms in the cell during division, and takes part in the distribution of chromatids to the daughter nuclei. The chromosomes become arranged around its equator. It appears to be a gel, largely composed of longitudinally arranged protein fibres which are synthesized shortly before division. The movement apart of the chromatids (or chromosomes) may be due to the contraction of some of the fibres.

SULPHONAMIDES   Sometimes called sulpha drugs. A group of organic compounds containing the group $SO_2 . NH_2$ as its derivatives. The first drug which could be used effectively to kill bacteria without killing the patient. First introduced in the 1930s and still widely used.

# Appendix 2

## Factors Affecting the Mutation Rate

### 1. Chemical Mutagens

A whole range of chemical substances will induce mutation, and are therefore said to have a *mutagenic* action, or to be *mutagens*. Such chemicals include hydrogen peroxide, urethane, manganous chloride, nitrous acid, mustard gas (dichloro-diethyl-sulphide) and many others. It may be that many of these mutagens act by causing replication faults or by changing the template. There is good evidence, too, that they may act to cause breaks in the DNA molecule. This can result either in the deletion of a whole segment, or in a re-arrangement of whole genes, a fact which can be demonstrated by working out the linear sequence of genes along a particular chromosome, thus making what is called a *linkage map*, before and after treatment with a mutagen.

**mutagen**

Some chemical mutagens can definitely be shown to act by causing chromosome breaks and other aberrations. Colchicine, for example, inhibits formation of the mitotic spindle and this causes the chromosome number to double at each division. Mustard gas can be shown to cause chromosome breaks in the fruit-fly *Drosophila*. These breaks are generally repaired immediately, but often with deletions, or with whole lengths of DNA moved to a different location. Such mutations should perhaps be termed 'chromosome mutations' to distinguish them from the gene mutations discussed elsewhere.

No evidence which has been offered so far proves that *every* change of a nucleotide produces a functional mutation; quite possibly it does not. It is worth remembering that there is not a one-to-one relationship between mutable sites on the DNA and any one amino acid of a protein. The *E. coli* enzyme *tryptophan synthetase*, which is the one referred to in the main text, illustrates this. It is possible to isolate a back-mutation in the tryptophan-requiring strain, in which the 'defective' amino acid is replaced by the correct one, restoring the activity of the enzyme. Investigation shows, however, that this is sometimes due not to a reversal of the original mutation, but to another change on an adjacent site. It is therefore not possible to say, in the present state of knowledge, to what extent every nucleotide will matter functionally.

**tryptophan synthetase**

The situation is further complicated by the fact that mutation may well involve many more than one nucleotide. On occasions, a mutation may involve the breaking and rejoining of whole chunks of DNA.

Simple substitutions of a nucleotide are likely to arise from time to time as errors in replication. As Unit 17 demonstrated, replication is a very accurate process; the probability of error in the incorporation of a new nucleotide may be as low as $10^{-8}$ or $10^{-9}$, but this is still appreciable. Such errors must arise through failure to form the proper hydrogen bonds, perhaps with the accidental pairing of adenine with guanine instead of thymine. These errors can be produced experimentally.

Quite a lot is now known about the effects of nitrous acid on the template. For example, it appears that it de-aminates adenine replacing $NH_2$ with keto groups, forming hypoxanthine. This will then pair with cytosine, rather than with the original thymine. Similarly, cytosine is de-aminated to uracil, which bonds to adenine instead of guanine.

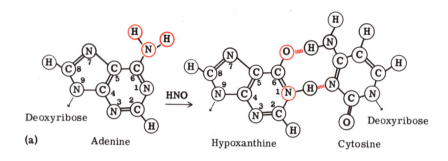

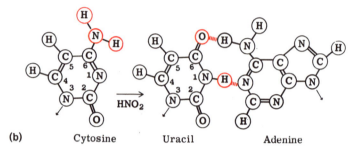

**Figure 20** *The effect of nitrous acid in causing amino-acid substitutions.*

## 2. Ionizing Radiation

Ionizing radiations (Unit 35), in this context meaning short wavelength electromagnetic radiation (e.g. X-rays and $\gamma$-rays) and accelerated particles with high energy (e.g. cosmic rays), have long been known to cause mutation. In 1927, Muller showed that the progeny of *Drosophila* irradiated with X-rays showed more mutations than progeny from non-irradiated parents (Table 7).

### Table 7

Muller's experiment in the induction of mutations by X-ray treatments in chromosomes of *Drosophila melanogaster*.

(The $T_4$ dose is twice as large as the $T_2$ dose.)

| Experiment | Number of chromosomes tested | Mutations observed | | |
|---|---|---|---|---|
| | | Lethals | Semi-lethals | Visibles |
| Control | 198 | 0 | 0 | 0 |
| X-rays ($T_2$) | 676 | 49 | 4 | 1 |
| X-rays ($T_4$) | 772 | 89 | 12 | 3 |

He later showed that the frequency of mutation is directly proportional to the dose of X-rays received, measured in *rads*. A rad is a unit of *absorbed radiation* (in tissues or any other material) (Fig. 21). It corresponds to 100 ergs/gramme of material.

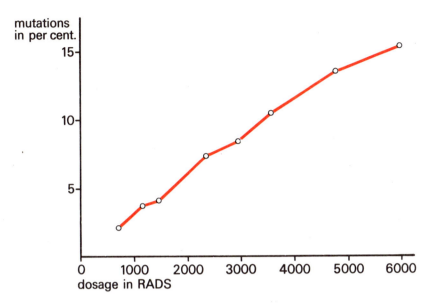

*Figure* 21   *The effect of X-rays on the mutation rate of* Drosophila.

It made no difference whether this dose was received over a few minutes at a high intensity, or over months or even years at a very low intensity; the number of mutations remained proportional to the total number of rads received. However, this simple relationship does not always hold true. Where the doses are large enough to damage the normal metabolic processes of the cell, it may make a difference whether the whole dose is given in one exposure or in several. Thus, in mice, it can be shown that a single dose of 1 000 rads produces fewer mutations than two doses of 500 rads given 24 hours apart. One of the reasons for this seems to be that where the single larger dose is given many of the mutated cells are killed and therefore do not come to light. There is also some evidence that the surviving cells from one dose are more mutable than the original cell population, when given another dose within certain time limits. In general terms, however, the larger the number of rads received, the greater the number of mutations produced, although there are complicating factors.

This obviously has great bearing in assessing the importance to humans of the radiation received from various sources, natural, medical, military or commercial. It is not true to say that there is a 'threshold' level of radiation below which additional mutations will not occur. We can only say that we are (or are not) prepared to put up with a particular avoidable increase of the mutation rate in the population.   **'threshold'**

It has also been shown that all radiation in this range has an equivalent effect if compared dose for dose. The mutagenic effect of such radiation does not seem to vary significantly over quite a wide temperature range, which is interesting in view of the effect of temperature on the spontaneous mutation rate (see below). Some other factors *do* vary the effect of a given number of rads, however, notably the amount of oxygen in the tissue. The lower the oxygen level the less frequent the mutations.

It is not certain exactly how the mutations are produced. Visible chromosome breaks may occur, in addition to changes within a gene; but it is not known whether the production of a mutagenic chemical (for example, hydrogen peroxide, which might account for the effect of the absence of oxygen mentioned above) follows a 'hit' close to a relevant hydrogen

bond, or whether the mechanism is even more direct. What is apparent is that it is a very localized occurrence. A particular dose or intensity of radiation does not produce a particular type of mutation throughout the tissue involved; presumably the damage occurs in a molecule only in the immediate vicinity of where radiation is absorbed and ionization occurs. In this context, it is interesting to note that ultra-violet light, which has a longer wavelength (of the order of thousands of ångströms), also produces mutations. However, different parts of the ultra-violet spectrum have vastly differing mutagenic abilities which precisely parallel the degree of absorption of the respective wavelength by nucleic acid. The higher the absorption, the greater the mutagenic effect.

It is, therefore, to be expected that ionizing radiations will cause a significant number of mutations in a population, whether the source is natural —e.g. cosmic rays, background radiation from granite rocks, etc.—or man made. It is doubtful, however, if this is the prime cause of 'natural' mutation. Certainly, in the case of *Drosophila*, it has been calculated that a fly would have to receive about 60 rads of radiation in four weeks (a fairly generous estimate of its generation time) to account for the observed rate of mutation, if this were the only cause. It is likely that in this time the fly would, on average, receive only about 0.05 rads, which would account for only 0.1 per cent of the observed mutations. However, insects are unusually resistant to radiation, so such calculations may not necessarily be relevant to mammals.

### 3. Temperature

Muller, in 1928, using the same technique as he employed to assess the effects of X-rays (see above) on mutation rate, also examined the effects of temperature.

Once again using *Drosophila*, he found that populations kept at higher temperatures showed a higher mutation rate. A rise in temperature from $19°$ C–$27°$ C ($8°$ C) increased the mutation rate by a factor of 2.5. Other experiments on different species have confirmed that a usual figure is an increase by a factor of 2–3 for a rise in temperature of $10°$ C.

This might be taken as an indication that multimolecular chemical reactions are involved in a normal mutation, as this is approximately the rate at which temperature rises effect such reaction rates. However, the increased mutation rate produced by irradiation does not vary with changes in body temperature, which suggests that such a reaction is not involved in the latter case. This is further evidence that 'spontaneous' mutations may not be caused by radiation.

# Appendix 3

## 1. Polyploidy and Variation

Another important source of genetic variation arises from a failure of normal division. This failure may come about if the spindle does not form correctly. The result of this is that one of the daughter cells gets all the chromosomes and the other none, leading to a diploid gamete and a zygote with one-and-a-half times the diploid number ('triploid'). If both gametes are diploid the zygote will be 'tetraploid'. Any organism which has more than the two normal diploid sets of chromosomes is said to be 'polyploid', so triploidy tetraploidy are two particular cases of polyploidy. Polyploidy is rare in animals, and usually results in a non-viable zygote. Thus the triploid human cell, a photograph of which is shown in Figure 22, is very unusual. However, one extra sex chromosome is not so uncommon, even in humans. Where the extra chromosomes do not form a complete set, the condition is not usually called polyploid, but it may nevertheless have a profound effect on the phenotype. For example, a human being with an additional chromosome of pair number 21 suffers from the condition known as mongoloid idiocy, or 'mongolism'. The chromosomes of one such person are illustrated here, also those of a 'feminized male' with one extra X chromosome, XXY instead of XY (male) or XX (female). In plants, however, polyploids are much more common—it is believed that about thirty per cent of all the flowering plants are polyploids. It used to be thought that this invariably lead to gigantism—for example, the giant sunflower is a polyploid variant of the normal one. However, although the individual cells generally *are* larger, the plant as a whole may not be, sometimes just some part of the plant useful to man may be enlarged, and this category includes a number of plants of great importance to man, such as some species of wheat, maize, tomatoes, potatoes, New World cottons and cultivated sugar cane. All these are polyploid versions of wild ancestors. For example, the wild American cotton has 26 chromosomes, 13 pairs, and is a normal diploid; but the form with the useful fibre is tetraploid, with 52 chromosomes forming 26 pairs. There is evidence to suggest that this has arisen from a cross between two different species of diploid cottons, forming what is generally called a *hybrid*. Hybrids are, of course, not necessarily polyploid; there are many diploid hybrids (see section 19.7), but these are normally sterile. Polyploid hybrid plants tend to show extra vigour and size: consider, for example, the difference between normal and giant sunflowers, wild wheats and oats and cultivated ones. Thus in plants polyploidy may constitute an important source of variation from an evolutionary point of view. Polyploidy may also occur at mitosis, if the chromosomes duplicate but the spindle fails to form and separate them. This effect can be produced in the laboratory by treating dividing cells with the substance colchicine. In multicellular animals, however, mitotic polyploidy is not very likely to be transmitted, and from this point of view is of less interest here.

**triploid**

**mongolism**

**hybrid**

### Relative importance of different sources of variation

We have considered the main sources of genetic variation in organisms: first, mutation, some of the causes of it, and its frequency; secondly, the effects of meiosis and sexual reproduction on the re-assortment of chromosomes and the re-combination of genes within the chromosomes; finally, a brief look at increases in the chromosome number, polyploidy. The relative importance of these factors in producing changed genotypes will

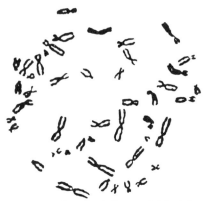

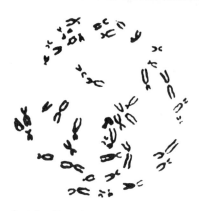

1 Chromosomes of normal Human Male (displayed in sequence below)

2 Same but for Female

| | | | | | |
|---|---|---|---|---|---|
| 1 | 2 | 3 | 4 | 5 | 6 |
| 7 | 8 | 9 | 10 | 11 | 12 |
| 13 | 14 | 15 | 16 | 17 | 18 |
| 19 | 20 | 21 | 22 | | X Y |

| | | | | | |
|---|---|---|---|---|---|
| 1 | 2 | 3 | 4 | 5 | 6 |
| 7 | 8 | 9 | 10 | 11 | 12 |
| 13 | 14 | 15 | 16 | 17 | 18 |
| 19 | 20 | 21 | 22 | | X X |

| | | | | | |
|---|---|---|---|---|---|
| 1 | 2 | 3 | 4 | 5 | 6 |
| 7 | 8 | 9 | 10 | 11 | 12 |
| 13 | 14 | 15 | 16 | 17 | 18 |
| 19 | 20 | 21 | 22 | | X X Y |

3  Human Triploid ( spontaneously aborted )

| | | | | | |
|---|---|---|---|---|---|
| 1 | 2 | 3 | 4 | 5 | 6 |
| 7 | 8 | 9 | 10 | 11 | 12 |
| 13 | 14 | 15 | 16 | 17 | 18 |
| 19 | 20 | 21 | 22 | | X X Y |

4 Feminised  Male ( Klinefelters syndrome, sterile )

| | | | | | |
|---|---|---|---|---|---|
| 1 | 2 | 3 | 4 | 5 | 6 |
| 7 | 8 | 9 | 10 | 11 | 12 |
| 13 | 14 | 15 | 16 | 17 | 18 |
| 19 | 20 | 21 | 22 | | X Y |

5 Mongoloid  Idiocy ( male )

| | | | | | |
|---|---|---|---|---|---|
| 1 | 2 | 3 | 4 | 5 | 6 |
| 7 | 8 | 9 | 10 | 11 | 12 |
| 13 | 14 | 15 | 16 | 17 | 18 |
| 19 | 20 | 21 | 22 | | 0 X |

6 Sexually immature Female ( Turner's syndrome, sterile )

*Figure 22    Photomicrographs of stained human chromosomes, as they appear in the cell, and displayed in sequence of homologous pairs. Asterisks mark the abnormal arrangements in 4, 5 and 6.*

vary enormously, and it is not worth generalizing. However, meiosis followed by fusion of the haploid gametes certainly leads to the most frequent changes in genotype—effectively every time in fact. These changes will, in principle, be small ones, because they are re-combinations of characters already being carried in the genotype of the species. But for some reason they should—once again in principle at least—all be viable characters. In contrast, mutation can produce quite new genes; but when an organism is an enormously complicated system of interrelated and interdependent structures and reactions (and most are), arbitrary changes are most likely to be harmful to a greater or lesser degree. When an undergraduate from a residential university celebrates, and afterwards replaces two of the semi-conductors in a computer with a tomato, or other randomly chosen component, it would be surprising to find that a useful change had been made; and though just the simple deletion of one small part might not affect the performance of the entire machine, it is likely to make it slightly less efficient, not more so. It will only be rarely that the operator will try out the change next morning and find that the computer works better than it did before.

It is difficult to compare polyploidy with the other changes, as it is so unclear why in some cases it results in 'more of the same', which is often good for the species, whilst in many organisms it merely results in genetic disaster. However, it is interesting to take note of the Evening Primrose, *Oenothera lamarckiana*, in this connection. It was as a result of studying this plant in 1901 that de Vries first coined the term 'mutation' in a genetic context, to describe the sudden changes he observed in the organism in his garden. It produced: a 'mutant' *Oenothera gigas*, of increased size; a dwarf form, *nanella*; several colour changes, and so on. Modern research has shown, however, that only about two of these new forms were caused by what *we* call mutants. The others were caused by a variety of genetic changes. *Gigas* is a straightforward tetraploid version of the parent plant (*lamarckiana*), having 28 chromosomes instead of 14. Others had 15 chromosomes, as one was present three times instead of twice. *Nanella* was the result of a complicated piece of genetic behaviour, which will not be elaborated upon here, stemming from the fact that *lamarckiana* is a type of hybrid, and it produces two very different types of gametes, which leads to curious results when it fertilizes itself. Thus the variations were being produced by mutation, polyploidy and an unusual consequence of meiotic assortment of chromosomes. In this particular case the three processes were of comparable importance.

## 2. Challenges to the Darwinian View

Even after the rise of genetics as a science had provided massive support for a modified Darwinian view, situations were demonstrated which some scientists considered to show support for a form of Larmarckian mechanism. Try to interpret the following examples according to each theory in turn, and see which you find the most convincing.

1 A Russian agriculturist called Lysenko, claimed during the 1930s to have established beyond all doubt the inheritance of acquired characters in crop plants, particularly wheats. In one of the best known of his experiments, he took a widely used variety of wheat, which was giving only a modest yield in the field, and bred it for several generations under test conditions, with good protection, water supply and large quantities of fertilizers, etc. The yields under these conditions rose sharply in successive generations. Then it was tried for a year under normal field conditions and gave a higher yield than it had originally. He claimed that by producing heavier yields under ideal conditions, he had actually 'conditioned' the wheat to transmit this ability to subsequent generations.

**Can you think of one explanation for his results in Darwinian terms—and how you would test your interpretations?**

Yet the Lysenko story is of more than merely passing interest, for it also provides an example of the relationship between science and wider issues. In the context of the agriculturally impoverished Soviet Union of the 1930s and 1940s, Lysenko's ideas were seized upon eagerly by those attempting to improve agricultural productivity. His opponents, classical geneticists led by the famous plant-breeder Vavilov, were forced into the defensive and many of them lost their jobs and some their lives. Lysenko's theories *had* to be right, it was claimed, and the 'normal' rules of scientific procedure and evidence were disregarded. The situation in the Soviet Union has since been 'normalized' once more, Lysenko's opponents have been rehabilitated—sometimes posthumously, while Lysenko himself continues to run a minor institute. The dispute divided the scientific world both in and out of Russia, for it was concerned not merely with the *facts* of the particular matter (and the complexity of the relationships between the organism, its development, genetics and environment are still only partially understood today)—but still more importantly with the relationship between the ideas and paradigms (Unit 1) of a scientist and the social environment in which he works. We return to some of these issues in Units 33 and 34.

2 In 1949, a curious transformation was shown to occur in the unicellular animal *Paramoecium aurelia*. If a suspension of some thousands of paramoecium is injected into the blood stream of a rabbit, the rabbit reacts as if these were germs (as it would with any foreign protein) and manufactures a substance in the blood which will destroy them. After this, samples of that rabbit's blood, if added to a culture of the paramoecium, will destroy it. However, it will only kill paramoecium of that particular strain. Other strains, say from another pond, will not be affected, unless the rabbit has been injected with them also. So we can

Under ideal conditions, with all the normal hazards removed, the heavy cropping strains would be favoured, as they produce more, and larger, seeds. Thus Lysenko may well have merely selected the genotypes which gave the heaviest yields per plant, disregarding any other possible selection pressures. When the wheat was first returned to the field, the predominance of the heavy cropping genotypes led to higher yields. These genotypes were not necessarily those giving high resistance to rust fungus, drought, high winds, etc., which is why they were not dominant in the original population. Thus, on Darwinian theory, he would have merely selected for those genotypes most suited to an ideal environment and *not* those best suited to the open field. On this basis, we should expect a very high mortality among these genotypes in the field, and an increase in the proportion of seeds derived from the smaller, hardier plants. This appears to have been the case; after one season the proportion of heavy cropping plants fell away, as did the total yields.

say that the different strains of *Paramoecium aurelia* fall into several different categories with respect to rabbit blood. The strains breed true with regard to these categories—the descendants of a culture in category 'A' are also in that category. However, if you take a culture of paramoecium in category 'A' and subject to a *weak* solution of the blood of a rabbit previously injected with that strain, so that the paramoecium are not actually killed, you find that the culture has changed category to one of the other groups; and it breeds true.

It appears that the nature of the substance produced by the rabbit is determined by bodies in the cytoplasm of the paramoecium. But in a paramoecium of, say, category 'A' not all the bodies are appropriate to this category; merely a great majority of them. It also contains different proportions of the other bodies, those which would, if in the majority, make it belong to one of the other categories. Thus, if a paramoecium in category 'A' is attacked by a weak solution of the blood of a rabbit containing the 'anti-A' substance, the 'A' bodies are destroyed. If the animal survives, one of the other groups of bodies multiplies up and becomes dominant; thus the animal has changed categories, and it will breed true for its new category. Does this seem to you to be a genuine case of the inheritance of an acquired change?

## Section 19.2

### Question 1 (*Objective 1*)

Which of the following (a–f) are true and which false?

1 Crossing over is:
  (a) the exchange of genetic material between homologous chromosomes;
  (b) the mixing of genetic material from two different species;
  (c) the process by which homologous pairs of chromosomes come to be randomly orientated during the first meiotic division.

2 Variation in the genotype will not be expected:
  (d) in a clone of micro-organisms;
  (e) between identical twin mammals;
  (f) in the offspring of a mating between identical twin mammals.

### Question 2 (*Objective 1*)

Which do you consider the clearest and most explanatory of the definitions of the gene below?

(a) A gene is the·smallest part of the DNA of a chromosome which will determine the structure of a complete polypeptide in the phenotype.

(b) A gene is a discrete particle or segment of the chromosome which determines a trait or character in the phenotype.

(c) 'What we shall call genes are special, separable and therefore independent "conditions", "bases" or "materials" that are present in the germ cells, and which determine at least many properties of the organism' (paraphrased from Johannsen, 1909).

## Sections 19.2 and 19.6

### Question 3 (*Objective 2*)

In various parts of this Unit you have been given information about, or asked to consider, various aspects of the disease sickle-cell anaemia. Now turn to the list of objectives and consider objectives 2 (a–g). Which of these objectives may be illustrated by what you know of sickle-cell anaemia, and why?

### Question 4 (*Objective 4*)

Potentially, sexual reproduction will be expected to confer certain advantages on a species, when compared to asexual reproduction. What is this advantage (a) and how does it arise (b)?

## Section 19.3

**Question 5** (*Objective 1*)

Select the best definition of phenotype:
(a) the arrangement of genetic material on which selection can act;
(b) the physical or 'bodily' expression of the genes or genotype;
(c) the structure resulting from the interaction of the physical expression of the genes or genotype with the environment.

## Section 19.5

**Question 6**

Write not more than half a dozen sentences which summarize what you consider to be the important points made in section 19.5.

## Section 19.6

**Question 7** (*Objective 2(e)*)

Write in not more than half a dozen sentences what you consider to be the important points made in section 19.6.

## Section 19.7

**Question 8** (*Objective 7*)

Join the ranks of the illustrious by attempting to define 'a Species'.

## Section 19.8

**Question 9** (*Objective 8 (c)*)

Draw up headings for an essay on the theme that 'Change in the human genotype is likely to be an important factor in future human evolution'. *Do not attempt this until you have listened to the radio programme of this Unit.*

## Question 1

(a) true; (b) false—this is hybrid formation; (c) false; (d) false, mutation may occur; (e) substantially true, as they are derived from a single egg. However, there may have been mutations in individual cells of the body, but these would probably be undetectable. (f) a non-starter. The parents are genetically identical, and therefore of the same sex, so they are unlikely to have any offspring.

## Question 2

(b) has the advantage of simplicity, but does not explain much. (a) is the most precise description, and in line with the thinking in this Foundation course (c) is the first definition, and still holds good, but suffers from the same disadvantages as (b).

## Question 3

2(c) The substitution of valine for glutamic acid in position six of the $\beta$ chain of the haemoglobin of the blood, a consequence of a single mutation.

2(e) Possessing the sickling gene in the heterozygous condition may be considered an adaptation to life in a malarious area.

2(f) The maintenance of the Hb.S gene in 20 per cent of the population in some malarious areas (but not elsewhere) provides a clear example of the effect of conflicting selection pressures.

2(g) It provides such an example both in its frequency in malarious areas, and its rapid elimination from groups which have migrated to non-malarious areas.

## Question 4

(a) It gives rise to greater *variation* than could be expected from asexual reproduction. This increase in the variety of genotypes should increase the ability of the species to adapt when acted on by differing selection pressures.

(b) 1  Because *two* individuals contribute genes to each offspring.
   2  Because in gamete formation there is rearrangement of gene combinations due to the independent assortment of the chromosomes and crossing over.

## Question 5

(c) is better than (b). The physical expression of the genotype is indeed the phenotype, but it is only a theoretical entity if divorced from the environment. *Some* environment or other *must* have interacted with it.

**Question 6**

1 The development of evolutionary theory in the eighteenth century is briefly examined, and its formulation on the basis of the inheritance of acquired characters is discussed.

2 An outline of Darwin's theory of evolution by natural selection is given, and the essential ingredients of it (Overproduction, Competition, Variation) are discussed.

3 The resistance to attack by phage shown by *E. coli* is used as an example of an experimental situation allowing predictions on Lamarckian and Darwinian theory to be tested.

**Question 7**

1 The meaning of 'fitness' in the genetic sense is defined, and the selection of strains of organisms resistant to damaging biological and chemical agents described.

2 This is illustrated by reference to *Staphylococcus* and penicillin resistance, houseflies and DDT resistance and the resistance of strains of the rat and rabbit to Warfarin and myxomatosis, respectively.

3 Other effects of selection on populations mentioned include their ability to adapt to exploit new food sources, and their ability to adapt to use the existing food resources more efficiently (as in the selection of faster-growing domestic animals).

4 The effect on the gene pool of a population of two conflicting selection pressures is exemplified by the distribution of the sickling gene.

5 The stabilizing effect of selection is mentioned.

**Question 8**

Beyond what we have said in the text, we can only help you by offering you three other definitions:

1 'Speciation occurs at that stage of the evolutionary process at which the once actually or potentially interbreeding population becomes segregated into two or more separated populations which are physiologically incapable of interbreeding'.

        (Adapted from Dobzhansky, 1941.)

2 'Species are groups of actually or potentially interbreeding populations which are reproductively isolated from other such groups.'

        (E. Mayr, 1942, *Systematics and the Origin of Species*.)

3 A species is a group of organisms able to inter-breed to produce fully viable offspring, or which may be presumed capable of doing so should circumstances have permitted, but which if given the opportunity to mate outside the group will not thereby produce descendants able to compete successfully with members of the parental group.

        (C. B. Goodhart, 1967.)

**Question 9**

In such an essay you must bear the following points in mind.

1 Human evolution has undoubtedly involved selective change of the genotype in the past.
2 Some changes due to selection can be *shown* to be occurring now.
3 We ourselves are changing our environment faster than we would expect obvious physical changes to arise by selection.
4 However, adaptive responses to our new environment may not necessarily *be* obvious physical ones. For example, selection for behavioural characteristics able to resist stress due to overcrowding, or for tolerance to particular contaminants in food or air, may be a very rapid process.
5 A combination of our well known technological resourcefulness and our less apparent wisdom may stabilize our environment completely, and remove the causes of competition between human beings. This could change our views on the effects of selection applied to variation as they affect man.

## Answers to In-text Questions

### Answer 1

Sixteen genotypes:

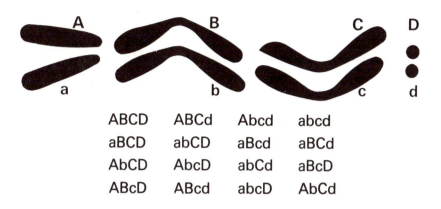

| ABCD | ABCd | Abcd | abcd |
| aBCD | abCD | aBcd | aBCd |
| AbCD | AbcD | abCd | aBcD |
| ABcD | ABcd | abcD | AbCd |

### Answer 2

There will be four different genotypes carried by the sperm for these characters:

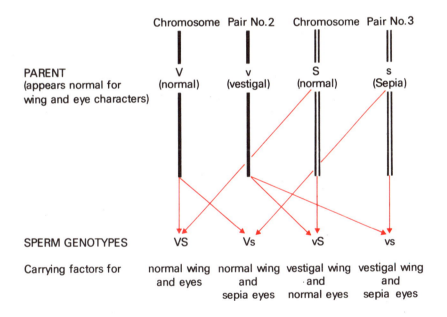

### Answer 3

Although there was a considerable overlap, the average size of the beans produced by group A were smaller than those from group B, though they were all grown under roughly similar conditions. Thus his selection was effective in separating out the two extremes of size, indicating that the difference had a genetic basis.

### Answer 4

Because all the group were of the same genotype. Therefore the size differences were only phenotypic, due to environmental differences.

## Answer 5

Yes, the genetic differences will remain.

## Answer 6

Attack by the phage in the Petri dishes induced a change in the relevant genes of a proportion of the population, thus conferring a heritable resistance on them.

## Answer 7

At some time in the course of the multiplication of the bacteria in the liquid cultures, spontaneous mutations occurred giving the possessors and their descendants immunity. These bacteria would be isolated when the whole population was attacked by phage.

## Answer 8

If the change in the genotype occurs in response to the phage attack, the numbers of resistant bacteria should be roughly the same in each Petri dish. As there are enormous numbers of bacteria in each dish, and the process of becoming resistant is presumably always the same, the variation between dishes should be very small.

If resistance is a result of random mutation you would expect the numbers of resistant bacteria to vary widely from dish to dish. For example, if one mutation occurred in the culture seven generations (say three hours) earlier, then by the time liquid is poured across the jelly there will be 64 mutants, each founding a resistant colony.

## Answer 9

Pair formation was unaffected by the appearance of the female, but much affected by that of the male.

## Answer 10

(a) Probably very relevant in view of the statement that 'it always pre-ceeded copulation', but these experiments do not provide evidence one way or the other.

(b) No information except in relation to pair selection.

(c) Very relevant, but the degree of relevance depended on (f).

(d) Irrelevant.

(e) Irrelevant.

(f) Relevant.

## Answer 11

Not very. The whole purpose of the work was to attempt an objective analysis of the behavioural relationships between the birds. To start with 'attachment' (affection?) as a hypothesis would seem unhelpful, particularly as it is hard to see how such a hypothesis could be tested.

## Answer 12

(a) No. It takes an anthropomorphic view (see Unit 1) of selection, and is teleological (see Unit 18).

(b) Selection will only favour isolating mechanisms in this way if the hybrids are less well adapted to the environment than the 'pure' species. Thus Smith is *assuming* this to be the case.

## Answer 13

Yes.

His results suggest that the markings and the birds' response to them, could well account for all or much of the apparent failure to interbreed. The work does not, of course, exclude the possibility of there being other, even stronger reasons, but it is difficult to see how these could be excluded, whilst necessarily working under natural conditions.

## Acknowledgements

Grateful acknowledgement is made to the following sources for illustrations used in this Unit:

A. C. ALLISON, Figs. 14, 15 and 16; BANTA TEACHING AIDS, Fig. 22; W. A. BENJAMIN INC., Figs. 2 and 20 in J. D. Watson, *Molecular Biology of the Gene*; J. & A. CHURCHILL LTD., Fig. 13 in G. E. W. Wolstenholme and M. O'Connor, *Bacterial Episomes and Plasmids*; SOCIETY FOR THE STUDY OF EVOLUTION, University of Kansas, Fig. 18.

The Open University

*Science Foundation Course Unit 20*

SPECIES AND POPULATIONS

*Prepared by the Science Foundation Course Team*

THE OPEN UNIVERSITY PRESS

# Contents

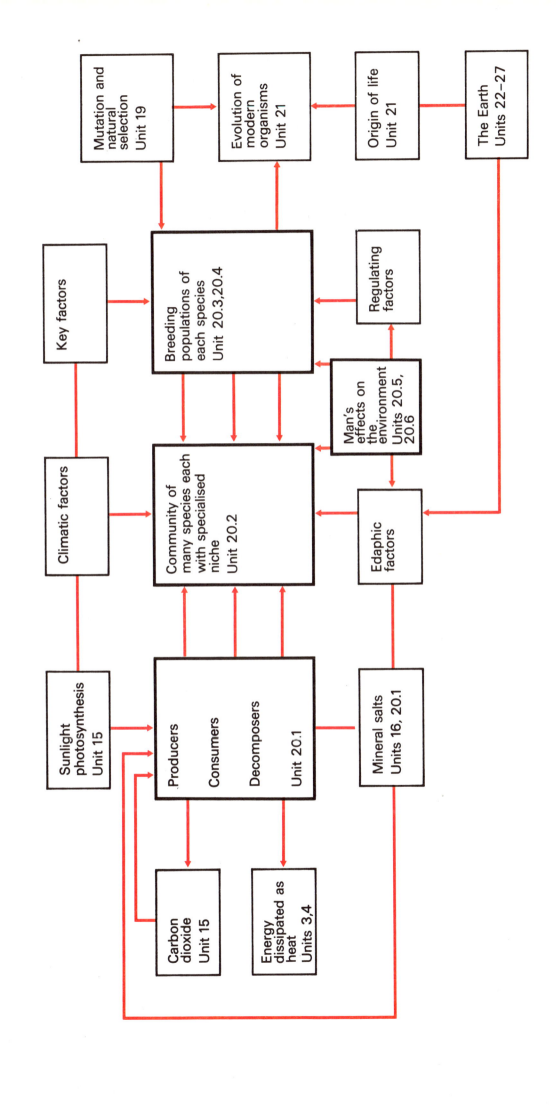

**Table A**

# List of Scientific Terms, Concepts and Principles used in Unit 20

| Taken as pre-requisites | | | Introduced in this Unit | | | |
|---|---|---|---|---|---|---|
| **1** | **2** | | **3** | | **4** | |
| Assumed from general knowledge | Introduced in a previous Unit | Unit No. | Developed in this Unit or in its set books | Page No. | Developed in a later Unit | Unit No. |
| general appearance of common types of British vegetation | species | 19 | community (as defined by ecologists) | 9 | adaptive radiation and specialized niches | 21 |
| | natural selection | 19 | carnivore | 11 | geochemical cycles | 25 |
| general habits of common British animals | vital activities | 18 | consumer (= heterotrophe) | 11 | hydrological cycles | 25 |
| | cellular metabolism and energy flow | 15, 16 | detritus feeder | 11 | special features of vertebrates, insects and molluscs | 21 |
| main climatic areas of the world | photosynthesis | 16 | herbivore | 11 | mammals of Australia | 21 |
| | protein synthesis | 15, 16, 17 | producer (= autotrophe) | 11 | climatic zones of the world | 24 |
| | | | production | 13 | nitrogen cycle | 34 |
| | autotrophes | 15 | ecological pyramids | 14 | pollution of the environment | 34 |
| | heterotrophes | 15 | food chains and webs | 14 | | |
| | aerobic respiration | 15 | parasite, specific parasite | 14 | | |
| | calorific value of food | 15 | energy flow through community | 15 | | |
| | sexual reproduction | 19 | decomposer | 17 | | |
| | exponential increase | 2, MAFS | cycling of carbon, nitrogen, sulphur and phosphorus | 18 | | |
| | | | niches | 21 | | |
| | | | climatic factors affecting plants and animals | 22 | | |
| | | | edaphic factors affecting plants | 24 | | |
| | | | competition between organisms | 25 | | |
| | | | mortality and mortality factors | 26 | | |
| | | | fecundity and birth rate | 27 | | |
| | | | survivorship curves | 28 | | |
| | | | logistic curve | 30 | | |
| | | | population cycles | 31 | | |
| | | | stability and change in communities | 34-36 | | |
| | | | territorial behaviour of birds | 34 | | |
| | | | key factor analysis | 35 | | |
| | | | density dependent factors | 37 | | |
| | | | regulating factors | 37 | | |
| | | | weeds and pests | 46 | | |
| | | | biological control | 48 | | |
| | | | effects of pesticides | 48 | | |
| | | | human population explosion | 49 | | |
| | | | exploitation of natural resources | 51, TV | | |
| | | | searching behaviour of parasites | 39, TV | | |

## Objectives

When you have completed the work for this Unit you should be able to:

1 Define correctly, or recognize the best definitions of, or distinguish between true and false statements concerning each of the terms and principles listed in Table A.
(Tested in *SAQs* 1, 5, 8, 12, 15)

2 Given relevant information about groups of organisms in a community, either recognize the position of each in given diagrams *or* construct from this information:
    (a) a food chain or web;
    (b) an ecological pyramid.
(Tested in *SAQ* 3 and Appendix 1)

3 Given relevant information about organisms to state the part they probably play in:
    (a) the carbon cycle;
    (b) the nitrogen cycle;
    (c) the sulphur or phosphorus cycles.
(Tested in *SAQ* 4)

4 Given figures in joules for two of the three following processes:
    (a) production;
    (b) respiration;
    (c) consumption of food;
to construct an energy flow diagram for the organisms concerned.
(Tested in *SAQs* 2, 3)

5 Given relevant information concerning:
    (a) a list of organisms;
    (b) types of environment;
to match organisms and environments.
(Tested in *SAQs* 6, 18)

6 Draw a population/time graph to show the effect of given changes in environmental conditions on the numbers of individuals in a population or interpret such graphs.
(Tested in *SAQ* 13 and Appendix 1)

7 Given graphs showing the relation between a given factor and numbers in a population, to recognize whether the factor is:
    (a) a key factor;
  or (b) a regulating factor.
(Tested in *SAQ* 13 and Appendix 1)

8 Given appropriate information, to calculate and construct mortality or survivorship curves.
(Tested in *SAQs* 9, 10, 11 and Appendix 1)

9 Choose the most plausible prediction from a given list, given information about changes within a simple food chain, or within a prey-parasite-hyperparasite chain or in a natural community. Suggest experiments to test this prediction.
(Tested in *SAQs* 17, 18, 19 and Appendix 1)

10 Select from a given list the factors which ought to be investigated before:

    (a) a given foreign species;

    (b) a given chemical substance,

is introduced on a large scale into an environment.

(Tested in *SAQ* 14)

11 Solve problems related to the geographical distribution of organisms using material drawn from this Unit and Unit 19.

(Tested in *SAQs* 6, 7)

12 Select from a matrix *or* recognize as true or false, statements about the following different types of behaviour:

    (a) territorial;

    (b) mating and courtship (from Unit 19);

    (c) searching (for food or host);

    (d) cryptic (concealment) (from Unit 19).

(Tested in *SAQ* 16)

13 Use the principles given in this Unit to make hypotheses or design experiments or draw conclusions from listed data not treated in the Unit.

(Tested in *SAQs* 17, 18 and 19 and Appendix 1)

## Students' Guide

No organism lives in complete isolation from other organisms. This Unit covers the study of organisms in their environments (the science of 'ecology') and illustrates some of the basic principles involved. We need to discuss species of plants and animals; some of these have precise English names but others have no common names or share names with other species. For the latter, we have given and used the 'Latin name', the internationally recognized name for the genus and species within the Linnean system of nomenclature described in Unit 19. Do not try to learn these names and do not worry if you do not know all the sorts of animals and plants that are mentioned in this Unit. If we did not use named examples the text would be vague and unsatisfactory, but the examples are chosen to illustrate general principles or methods of analysis and it is these principles and methods that are important. Try not to lose sight of the wood for the trees!

There are two prescribed texts for this Unit. You will need to have both these texts beside you while you read, since there are references to figures, tables and text in both books.

There is an order and a pattern underlying the distribution and numbers of plants and animals. In 20.1, we show how communities of organisms are linked together by their energy requirements and need for certain chemical compounds. Part of this section is a commentary on one of the prescribed texts for the Unit. Through the study of 'production ecology', we gain an insight into the relative numbers of different sorts of organisms and some of the factors that limit the size and complexity of communities.

In 20.2, the structure of communities is approached from a different point of view, that of the parts played by the many species that compose them. Each species occupies a 'niche' (defined in 20.2); comparable niches may be occupied in different parts of the world by different species. We study some of the factors controlling the distribution of species through the world.

Quantitative studies of populations depend on collecting and counting samples and then analysing the numbers obtained. In 20.3, we consider how population numbers change as organisms grow older and how populations are affected by birth rates and death rates.

Studies of numbers of individuals in populations show that some species are rare and others are common but for both types of species there are fluctuations; some of these are regular 'cycles', but others are irregular and, at present, unpredictable.

'Population dynamics' is the study of population numbers and how and why they vary; this is the subject of 20.4. If sufficient relevant information is available, then it is possible to identify the causes of death that are responsible for the fluctuations and also to identify other causes of death that vary in such a way that the population numbers tend to return towards the average values. If you are having difficulty, then you should read rapidly through this section without attempting to work out the figures for yourself.

Appendix 1 (Black) is a structured exercise to give you practice in applying methods and principles mentioned in sections 20.1, 20.3 and 20.4.

Communities are generally close to a state of balance under natural conditions but change slowly with time. Interference with a community is likely to lead to rapid changes—in our present state of knowledge, some of these changes may be unexpected and drastic. Some of the effects of the sudden reduction in numbers of rabbits after the onset of myxomatosis in Britain are discussed in 20.5.1.

The impact of man on the environment is currently receiving much publicity. In 20.6 we consider some of the biological implications of modern agricultural methods, particularly in relation to weeds and pests. The human population explosion is creating serious problems, discussed in 20.6 from the point of view of production ecology. Human survival may depend on rational exploitation of food resources, some of which are at present over-exploited, as illustrated in the TV programme.

The study of ecology, of the principles underlying the complex patterns of species distribution and community structure through the world, is of fundamental importance for human survival. Here in Unit 20 you meet some of the modern ideas and methods of this fast-growing branch of biology.

In Unit 19 you learnt what biologists mean by the term 'species' and that they believe that modern species have evolved by modification from pre-existing species. The most probable mechanism for this evolution is natural selection which acts on a group of individuals that form an interbreeding population. The genes of these individuals form part of a 'gene pool' on which selection acts so that at any time the 'wild' population will be well adapted to the environment in which it lives. Since individuals that are less well adapted are less likely to survive than the others, the 'survival of the fittest' means competition among individuals of the same species—a form of intra-specific competition.

In this Unit we are more concerned with inter-specific relationships, with the interactions between the many species of organisms living in the world today. The study of organisms in their environments—environmental biology—is called *ecology* (from the Greek for 'house'). At present there is considerable public interest in and awareness of the effects of man's activities on the environment and so on other species of organisms. Some special examples of this are included in Units 10 and 34. In this Unit we shall study the basic principles of ecology—the principles which govern the distribution and sizes of populations of organisms throughout the world.

*ecology*

As with most sciences, the early development of ecology was as a descriptive science, based on the compilation of lists of organisms present or absent from various places. Naturally, it is still important to know the distribution of different organisms and to understand the factors controlling this; these are discussed in 20.2. When species are listed for different areas, it is apparent that they form non-random associations which are called *communities*. These are formed of populations of different species found together in a definite area. Each community exists in a non-living environment made up of the medium (earth or water, or both) and the local climate. The community studied with its environment makes up an *ecosystem*. Communities and ecosystems can be studied at different levels—for instance, the community of an oak-wood includes a large number of smaller communities such as those of the oak-trees, the hazel shrubs, the herbs under the trees, the litter of fallen leaves and so on.

*communities*

*ecosystem*

The oak-wood itself is part of a terrestrial ecosystem, probably a lowland area of Britain, which in turn is part of the island ecosystem of Great Britain. Even this island ecosystem is not a unit independent of all others, since birds and butterflies and other less conspicuous organisms, as well as people, can move in and out from neighbouring continental areas. Thus it is necessary to define communities carefully when discussing and comparing them.

Ecologists have moved from the purely descriptive approach to two main lines of quantitative research. The first is the study of the structure of communities in terms of the throughput of energy and chemical compounds. This is called *production ecology* and we shall discuss its principles in section 20.1. The second quantitative approach is the study of the numbers of organisms of different species and how they vary in time and space and why they are common or rare. This is called *population dynamics* and it forms the content of sections 20.3 and 20.4.

Of course, all these types of approach are necessary for a complete understanding of the distribution of species of organisms in the world and of the size of their populations—an ideal far from achieved at present. The better scientists understand ecological principles, the better able they will be to predict the results of interference with existing communities by man or by any other agency. In sections 20.5 and 20.6 we shall discuss some examples of such interference.


## Reading


There are two prescribed texts:

John Phillipson, *Ecological Energetics*. Arnold, 1966.
Maurice E. Solomon, *Population Dynamics*. Arnold, 1969.

You will be directed to read certain sections and to look at certain figures and tables in both books. If you wish, you may read the whole of both books, treating them as black-page appendices.

If you wish to read one of the recommended books, you will find the following chapters are relevant to this Unit:

N. J. Berrill, *Biology in Action*. Heinemann, 1967. (Chapters 32 and 33.)
S. D. Gerking, *Biological Systems*. Saunders, 1969. (Chapters 5, 9 and 22.)
P. B. Weisz, *The Science of Biology*. McGraw-Hill, 1967. (Chapters 8, 9, 10 and 17.)
Rachel Carson, *Silent Spring*. Penguin, 1965. (This book was written in 1962 and started the concern for the environment that has increasingly made an impact in Europe and USA.)
K. Mellanby, *Pesticides and Pollution*. Fontana, 1969.

# .1 Production Ecology

Although production ecology is a more recent development than descriptive ecology, it provides a logical basis for the structure of communities, so it is sensible to study it first. A real understanding of the processes which determine the level of production is also essential for rational exploitation of world resources of food for the increasing human population—we shall refer to this again in 20.6.

The vital activities of all organisms (refer to Unit 18, section 2.1, if you need to remind yourself about them) depend on 'the use of energy'. As you read this text, you are moving your eyes, you are breathing, your heart is beating, your guts are digesting and assimilating food and nerve cells in your brain are active. As you read in Units 15 and 16, all this activity depends on biochemical processes in your body cells. These processes stop in dead organisms. One way of dying is through starvation, i.e. lack of raw materials essential for the biochemical processes. Thus life depends on the supply and utilization of certain raw materials and we can think of these as the source of energy for life processes (vital activities).

Photosynthesis is characteristic of plants. Summarize the main features of this method of feeding by answering the following questions:

**autotrophes**

1 **What are the raw materials needed by the plant?**
2 **What does the plant synthesize?**
3 **Where does the energy for photosynthesis come from?**

1 Carbon dioxide and water.
2 Carbohydrates—sugars which may then be converted into storage products such as starch.
3 Sunlight—energy is made available in the chloroplasts in the plant cells (usually in leaves of land plants).

The plant grows by producing more plant cells; proteins are essential components of cells.

**Beyond those already mentioned, what substances do plants require for the synthesis of proteins?**

Compounds containing nitrogen, phosphorus and sulphur. Plants use inorganic salts—nitrates, phosphates and sulphates—and these are often collectively called 'mineral salts'.

Since plants can synthesize their proteins entirely from inorganic substances—water, carbon dioxide, mineral salts—if provided with the appropriate wavelengths of light, they are called *autotrophes* (from the Greek for 'self' and 'food'). Organisms which cannot synthesize proteins from inorganic substances, but must obtain their amino acids by digestion of proteins manufactured by some other organism, are called *heterotrophes* (from the Greek for 'different' and 'food'). Typically these organisms cannot synthesize sugars either, and must obtain these from some other organism. Animals are heterotrophes; so are some plants such as the fungi (mushroom, moulds). So the growth of animals depends on them eating other organisms, living or dead. If they eat plants, as cows do, the animals are called *herbivores* (from the Latin for 'grass' and 'devour'); if, like lions, they eat other animals, they are called *carnivores* (from the Latin for 'flesh' and 'devour'). Animals, such as earthworms, that feed on fragments of decaying plant and animal material are called *detritus feeders* (from the Latin for 'wearing away').

**herbivores**

**carnivores**

**detritus feeders**

11

The release of energy within cells and the uses made of this energy were discussed in Units 15 and 16. There are some organisms, such as yeasts and various bacteria, that have unusual metabolic pathways, but most cells and organisms depend on aerobic respiration. Ultimately, they depend on absorbing oxygen from the environment; the uptake of this can be measured, often very accurately. Glucose is the substrate that is usually metabolized during tissue respiration; fats and proteins are metabolized in some organisms and under certain conditions, but these are much less common substrates than glucose. Thus it is generally reasonable to deduce the total output of energy of an organism or group of organisms by measuring the amount of oxygen absorbed from the environment and calculating how many joules* would be available were this oxygen all used in the tissue respiration of glucose. From respiration of 180 g of glucose, $283 \times 10^4$ J would be made available for the organism's activities.

respiration and energy

Heterotrophes (animals) obtain all their energy from their food. Often it is possible to measure the amount of food that is eaten; the energy thus taken into the animal can be calculated from the calorific value of the food (refer to Unit 15 for more information about this). If the food intake and oxygen uptake are known, an energy balance can be set up:

Calorific value of food = output of energy calculated from + X joules
(in joules)                the amount of oxygen absorbed
                           (in joules)

food and energy

If $X$ is a positive value, then the organism is able to grow or to lay down food reserves (e.g. fat deposits) or to produce reproductive bodies (e.g. eggs and sperm). But if $X$ is negative, the organism may be using up food reserves or wasting away.

With autotrophes (green plants), the measurement of oxygen absorbed in respiration is complicated because the plants perform photosynthesis.

**Show this by writing two equations:**
**(a) to summarize the process of respiration, in which glucose is converted to $CO_2$ and water;**
**(b) to summarize photosynthesis, generating glucose from $CO_2$ and water.**

(a) $C_6H_{12}O_6 + 6O_2 \rightarrow 6CO_2 + 6H_2O$
(b) $6CO_2 + 6H_2O \rightarrow C_6H_{12}O_6 + 6O_2$

**Suggest how (a) respiration and (b) photosynthesis could be measured for plants.**

In practice, similar plants or collections of plants are set up in closed vessels, one exposed to light and the other in a black, light-proof container. The change in amount of oxygen in the two containers is measured:

Plants can only perform photosynthe in light but they must respire both in light and in the dark. So it is possible measure respiration as the uptake of oxygen by a plant in the dark. To measure photosynthesis, the plant mu be in light. It will produce more oxygen than it uses in respiration but a correction must be made, adding the volume of oxygen consumed in respiration.

**How is the total amount of oxygen produced by photosynthesis obtained?**

By adding the amount of oxygen used up in the dark to the amount of oxygen produced in the light (after suitable adjustments for amount of material in the two containers). The plants in the light respire as well as perform photosynthesis; it is assumed that they respire at the same rate as similar plants in the dark.

## 20.1.1  Ecological energetics

This section is concerned with energy 'budgets'. The relationship between organisms and their environment in terms of the acquisition and dissipation of energy is important to ecologists interested in natural communities;

* *You will find that many books and articles* including your prescribed text *use calories or kilocalories (kcal) as units for energy instead of joules (the SI unit). To convert calories or kilocalories into joules, multiply by* 4.2 *or* $4.2 \times 10^3$ *respectively.*

it is also very important for the rational exploitation of world food resources, as we shall show in 20.6.

While studying this section, you will be referred to your prescribed text, *Ecological Energetics*. You will be told to read the following sections of that book: 1.6, 2.2, 2.4, 3.2 and 3.3 (adding up to about 4 000 words). Before you start to read each section, you should read the commentary on it in this text. Definitions of unfamiliar names and terms are given here. Note that the unit for energy in the book is the calorie or the kilocalorie (equivalent to 4.2 or to $4.2 \times 10^3$ joules).

DO NOT ATTEMPT TO MEMORIZE THE FIGURES OR NAMES.

### Section 1.6   *Energy transformations in nature*

Start reading on page 5, section 1.6, 'Energy transformations in nature'. Note the term 'production'—this is only loosely equivalent to the word 'productivity' as used in ordinary speech. Production is a measure of the total amount of new living matter formed per unit time; even if some part of this matter dies (or is consumed by some other organisms) during the time interval, it must still be included in the gross production. Since net production is gross production minus energy released in respiration, the matter which has died or been consumed by some other organism is also included in 'net production' for that time. You will meet examples of this later.

production

### *Definitions*

*incident energy*—the energy of the sunlight falling upon a particular place, including that of all wavelengths, visible and invisible (see Unit 2).

*efficiency*—is usually measured by expressing the amount of energy utilized by the particular organism being studied as a percentage of the total energy available to it. Efficiency is never 100 per cent!

*chemical energy*—calorific value.

*sensible heat*—heat that can be felt by man.

*perennial grass-herb vegetation of an old-field community*—the grasses and small flowering plants in a field which has gone out of cultivation.

*exudate*—secretion produced by plants, such as nectar and sap flows.

*peat deposition*—plant matter decays very slowly under certain conditions and the decaying plants accumulate to give a brown or black soil called peat (dried, this is highly combustible and still used widely as fuel in parts of Ireland).

*faeces*—these are the remains of food taken into the body but not assimilated.

*excretory products*—these are waste substances produced within the body and then discharged, e.g. urine and sweat. The term 'excreta' in ordinary speech usually includes faeces but, scientifically, faeces and excreta are different in origin since the former have never been assimilated into the body whereas the latter are by-products of metabolism.

*Now read section 1.6; then return to this text.*

### Section 2.2   *Food webs*

When a plant is eaten by one animal which in its turn, is eaten by another, the sequence of events can be expressed thus:

Plant → Herbivore → Carnivore.

This is called a *food chain*. You should be able to think of many possible food chains—here are three examples:

Lettuce → Snail → Thrush;

Grass → Vole (small rodent related to mouse) → tawny owl;

Grass → Sheep → Man.

Phillipson gives other examples (using the scientific names) in section 2.1. He then points out, in section 2.2, that the food chains in natural communities branch and overlap with each other to give complex patterns, called *food webs*. As an example, he considers the feeding relationships of the herring and illustrates this in Figure 2.1. All the organisms mentioned in the text are shown in the figure. Note that the arrows show the direction of feeding, *not* the direction of energy flow (as is the convention in the food chains above and the one on page 10 of the book). The herring eats sand eels, so energy from sand eels flows to the herring. In Figure 2.1, the organisms at the bottom of the diagram (diatoms and flagellates) are small floating plants. If you read Appendix 1 of Unit 18, you will have met them already. The organisms that feed on this *phytoplankton* (this is the term for small plants that live in the upper layers of the sea and lakes and drift with water currents) are called *zooplankton*. This means small animals that live in the upper layers of the sea and lakes; some of them are sometimes called 'water fleas'. The layer above the zooplankton in the diagram includes one fish, the sand eel, and three other unfamiliar organisms that are small but predatory.

food chains and webs

*Now read section 2.2; then return to this text.*

### Section 2.4 *Ecological pyramids*

The idea of constructing ecological pyramids proved a very helpful step forward in quantifying ecological observations, so it is important that you should understand about the different sorts of pyramid. Each type has its uses and its limitations; these are pointed out in this section of the book.

pyramids of numbers

*Figure* 2.2   Do not try to identify the organisms, but simply look at this as an example of a complicated pattern—in this case, based on the community of a small stony stream.

*Figure* 2.3   *Parasites* are organisms which obtain all their food from another organism, the host, which is usually much larger than its parasite. Usually the parasite lives on or in its host for most of its life. *Hyperparasites* are parasites of parasites—and therefore smaller than their hosts, the parasites of a larger host. Plant parasites include aphids (greenfly and blackfly) and other 'plant-bugs'; these insects may have roundworms as parasites inside them; they probably also have bacterial parasites.

*Figure* 2.4   Phytoplankton and zooplankton have already been defined when discussing the food of the herring.

The dry weight is obtained by heating the organisms in an oven at about 80° C until there is no further change in weight. All the water should then have evaporated and all other components of the body should remain and be weighed. The bodies of most organisms are about 60 per cent water; some organisms, such as jellyfish, have more than 99 per cent of water in their wet weight.

*Text below this figure*   Algae are simple plants. The group includes phytoplankton, such as the diatoms; these are the microscopic forms referred to here.

*Figure* 2.5  P=producers; H=herbivores; C=carnivores; T.C.=top carnivores. Look at the shape of the pyramid; ignore the words and figures. The data used for this diagram are shown in a different form in Figure 3-2 (see later).

*Text below this figure*  *Ensis* is a razor shell, quite common on some sandy beaches where it burrows near the low tide level. *Calanus* is shown in Figure 2.1; it is a water flea that is a very important part of the diet of the herring.

*Now read section 2.4; then return to this text.*

### Section 3.2  *Energy flow models*

The pyramid in Figure 2.5 shows the amount of energy utilized in one year by the organisms on or over one square metre of a spring-fed river in Florida. The rate of energy utilization of the different trophic levels (producers, herbivores, carnivores, etc.) can be expressed in other ways; one is shown in Figure 3.2. This is called an energy flow diagram because it indicates the sources from which each trophic level draws its energy and the actual fate of that energy; thus the complete diagram follows the fate of all the energy entering that area and available to organisms.

**energy flow diagrams**

*Look at Figure* 3.2 (*page* 28).

At the left is the amount of energy entering the system—the energy in the light absorbed by the plants in the spring. As a result of photosynthesis, about $21 \times 10^3$ kcal/m²/yr of plant material is produced (from $410 \times 10^3$ kcal of energy absorbed by the plants). The plants respire, using about $12 \times 10^3$ kcal, so that about $9 \times 10^3$ kcal of plant material is available as potential food (and source of energy) for herbivores. But some of this plant production leaves the area (as algae or bits of flowering plants washed downstream) and some dies and so is only available to organisms called *decomposers*. So the actual amount of energy as plant material available to herbivores is $3.37 \times 10^3$ (with $5.95 \times 10^3$ going to decomposers or to export and $0.49 \times 10^3$ entering the spring as imported plant material).

Following the diagram to the right, you can see that the potential food energy available to first level carnivores is only 383 kcal (out of the $3.37 \times 10^3$ kcal available to herbivores).

**How much energy is dissipated as heat, as a result of respiration of the herbivores?**

1 890 kcal—this is shown in the stippled area leading from the rectangle labelled 'Herbivores' to the block labelled 'Total heat.'

The production of first level carnivores is only a fraction of the production of herbivores (383 kcal compared with 3 368 kcal). The first level carnivores respire and are also available for export and for decomposers and as food for production of top carnivores. Look at the diagram and answer the following questions.

**What proportion of the energy available as food to first level carnivores is dissipated as heat through respiration? What proportion goes into the production of top carnivores?**

316/383 kcal is dissipated as heat during respiration and 21/383 kcal goes into production of top carnivores, representing about 83 per cent and about 5.5 per cent respectively of the food energy.

15

Look now at the block labelled 'Decomposers'. Compare the energy of the material available as food to the decomposers with the dissipation of heat as a result of their respiration. The food energy for decomposers is $5.06 \times 10^3$ kcal and the heat dissipated is $4.6 \times 10^3$ kcal (about 91 per cent of the food energy). Notice that the decomposers provide food for more decomposers, but this is only about 9 per cent of the total food of decomposers ($0.46 \times 10^3$ compared with $5.06 \times 10^3$ kcal).

Finally, compare the total heat dissipated as a result of respiration of plants ($12 \times 10^3$ kcal), herbivores and carnivores ($1.9 + 0.3 + 0.13 = 2.33 \times 10^3$ kcal) and decomposers ($4.6 \times 10^3$ kcal) with the gross production of plants. The total heat dissipated through respiration of the whole community is $18.93 \times 10^3$ kcal and the gross production of plants is $20.8 \times 10^3$ kcal—the difference is explained by the figures for export and import of material. But remember that the gross production of plants represents only a fraction of the energy in the light absorbed ($21/410 \times 10^3$ kcal or just over 5 per cent). The other 95 per cent of the incident light energy is not utilized by the plants and so is not available to any of the animals.

*Now look at Figure* 3.1 (*page* 25).

This is an energy flow diagram for another freshwater spring in USA. Ignore the figures inside the boxes (which indicate changes in standing crops). The other figures, mostly associated with arrows, are comparable with those in Figure 3.2. The stippled areas represent energy dissipated as respiration. There is a misprint—the respiration of micro-organisms should be 295 kcal, not 2.95. Notice that there is no figure for the input of energy as light. Plants are represented as a small box labelled 'algae' from which 655 kcal are available as potential food for herbivores. The total energy available for production of herbivores is given as 2 300 kcal:

**Where does most of this come from?**

It comes from 'Debris' 2 350 kcal less 705 kcal flowing to 'Deposit' and 'Micro-organisms'.

If you find this confusing, you are justified in doing so! The category 'Herbivore' in this diagram actually consists largely of organisms that are called 'decomposers' in Figure 3.2. These are animals that eat debris of plant origin; some also eat living plants such as algae attached to stones or to sticks, so that it is difficult to separate 'pure' herbivores from animals with wider feeding habits.

The term *herbivore* is used for the primary consumers in Figures 3.1 and 3.2. That these feed on plant debris in one, and on live plants in another, is a real difference between the two communities; this difference is discussed in the first paragraph of section 3.2. *Detritus* is the term for dead organisms in various states of breakdown; it usually consists largely of sticks and leaves.

When you look at Figure 3.3, to compare the grazing food chain with the detritus food chain (these are explained in the first paragraph), the figures you should compare are the following. For the marine bay, the food energy available from the phytoplankton (the producers) is 8 kcal; of this, 6 kcal pass (as living algae) along the grazing food chain and 2 kcal (as dead algae) along the detritus food chain. For the forest, the food energy available as leaves (from the producers, the trees) is 12 kcal; of this, 2 kcal pass (as green leaves on the tree) along the grazing food chain, while 10 kcal pass (as dead leaves forming 'litter' on the ground) along the detritus food chain.

respiration and production

grazing and detritus food chains

Do not worry about the other details of these two diagrams.

*Now read section 3.2; then turn back to this text.*

### Section 3.3    Decomposers

Comparison of the four energy flow diagrams referred to in section 3.2 has revealed the importance of dead organisms as a source of food energy. You may have wondered what sorts of organisms consume detritus and corpses. These decomposers are discussed in section 3.3, where experimental evidence is presented to show the importance of larger detritus feeders in enabling bacteria and other micro-organisms to attack dead leaves. Do not worry if you do not know all the organisms mentioned in the text—the point is to realize the complementary effects of the larger and smaller organisms and that both are important in the energy budgets of communities.

**decomposers**

*Now read section 3.3; then return to this text.*

We shall refer to some figures and tables in *Ecological Energetics* later in the text. You may read the whole book, if you wish, treating the rest of it as equivalent to a black-page Appendix.

### 1.2    Carbon, nitrogen and phosphorus cycles

In production ecology, as you have read in *Ecological Energetics*, 'energy' is traced from the sun, through plants, herbivores and carnivores of the grazing food chain and their dead bodies, and then through the detritus food chain. At each link of the chain, energy is 'dissipated' as a result of respiration and ceases to be available for the activities of organisms. Only part of the energy represented by plant production is 'fixed' by animals as production of herbivores; only part of herbivore production is transformed into carnivore production and so on. Without plants performing photosynthesis, production at all other trophic levels would eventually stop, as illustrated in Figure 1. 'All flesh is grass' is a true statement if it is read as 'all other living organisms are ultimately dependent for survival on autotrophes and especially on photosynthetic autotrophes'!

Key:

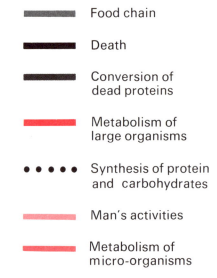

|  |  |
|---|---|
| ▬▬▬ | Food chain |
| ▬▬▬ | Death |
| ▬▬▬ | Conversion of dead proteins |
| ▬▬▬ | Metabolism of large organisms |
| ••••• | Synthesis of protein and carbohydrates |
| ▬▬▬ | Man's activities |
| ▬▬▬ | Metabolism of micro-organisms |

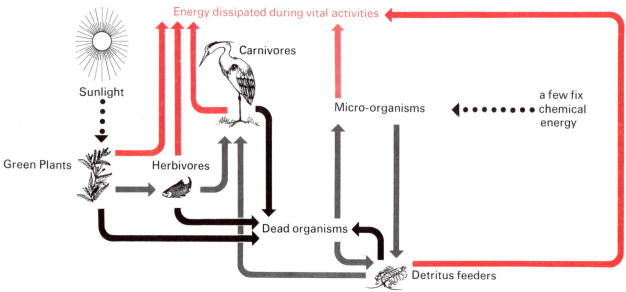

*Figure 1    Diagram to illustrate the 'flow' of energy through a community of organisms.*

Food webs are concerned with more than the 'flow' of energy from plant to herbivore to carnivore and so on. Plants synthesize proteins from simple raw materials—in the case of green plants, from water, carbon dioxide and mineral salts; herbivores obtain all their amino acids from the proteins of producers; carnivores obtain theirs from the proteins of herbivores.

**Do you expect the elements carbon, hydrogen, oxygen, nitrogen, phosphorus and sulphur (the constituents of proteins) to be dissipated and 'lost' to living organisms as energy is?**

**Or would you expect these elements to be 'cycled' and utilized again by living organisms?**

In fact, they are 'cycled'. Figures 2, 3 and 4 show the fates of carbon, nitrogen and phosphorus. Sulphur cycles in much the same way as phosphorus. Hydrogen and oxygen cycles are more complex, so we will not describe them here.

Micro-organisms convert proteins of dead organisms through various intermediates into carbon dioxide, water and inorganic salts (usually referred to as 'mineral salts').

Examine the Figures 2, 3 and 4, then answer the following questions:

**Are the following types of organisms essential for the complete cycling of (a) carbon, (b) nitrogen, (c) phosphorus?**

| | |
|---|---|
| **1 green plants** | **2 micro-organisms** |
| **3 herbivores** | **4 carnivores** |
| **5 detritus feeders** | |

Key:

| | |
|---|---|
| ▬▬▬ | Food chain |
| ▬▬▬ | Death |
| ▬▬▬ | Conversion of dead proteins |
| ▬▬▬ | Metabolism of large organisms |
| ••••• | Synthesis of protein and carbohydrates |
| ▬▬▬ | Man's activities |
| ▬▬▬ | Metabolism of micro-organisms |

micro-organisms

1 (a), (b), (c): Yes.

2 (a): No. (b), (c): Yes, unless man synthesizes massive amounts of fertilizer at enormous energy cost.

3 (a), (b), (c): No—plants in theory could survive without any animals being present, but they would be strictly limited by the amount of $CO_2$ present, since only plants and micro-organisms would be alive and respiring. Carnivores are strictly dependent on herbivores being present. Detritus feeders could survive on plant detritus only.

4 (a), (b), (c): No.

5 (a), (b), (c): No—in theory the micro-organisms of decay could cope with dead plants and animals without the intervention of detritus feeders. But, as you read in *Ecological Energetics*, p. 31, decay proceeds much faster if detritus feeders are present.

The presence in most communities of herbivores, carnivores and detritus feeders in addition to plants and micro-organisms results in a faster turnover of the essential elements carbon, nitrogen, sulphur and phosphorus. This is because the respiration of the animals yields $CO_2$, their excretion yields principally nitrogenous compounds, and their activities produce dead organisms and assist in their physical destruction. So the total production of a community, given that all other factors are equal, is likely to be greater where the food web is more complex than where it is very simple with few links.

## 20.1.3 Summary of section 20.1

Production ecology is the study of communities of organisms in terms of their energy budgets and of the cycling of essential elements. These impose a structure on the community such that the range of feeding habits and the numbers and biomass of each species must lie within limits related to the ultimate source of energy (usually sunlight). Inter-relationships between the species can be expressed as food webs; relationships between organisms of different trophic levels can be expressed as pyramids. Energy flow diagrams and pyramids of energy flow express certain relationships between organisms that we shall use later when discussing natural resources and human food supplies.

*Now you can do* SAQ 1 *to* 4.

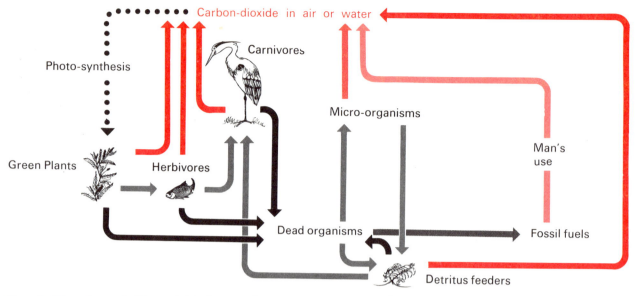

*Figure 2* The carbon cycle. The symbols/colours are the same as in Figure 1.

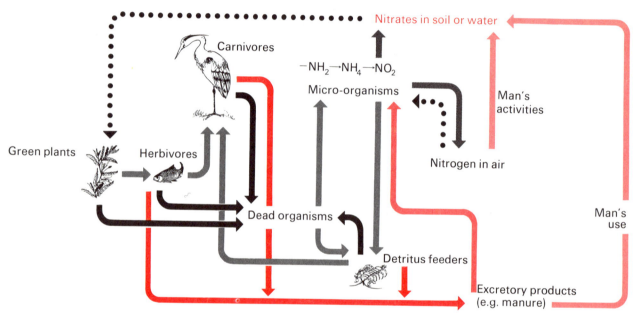

*Figure 3* The nitrogen cycle. The symbols/colours are the same as in Figure 1.

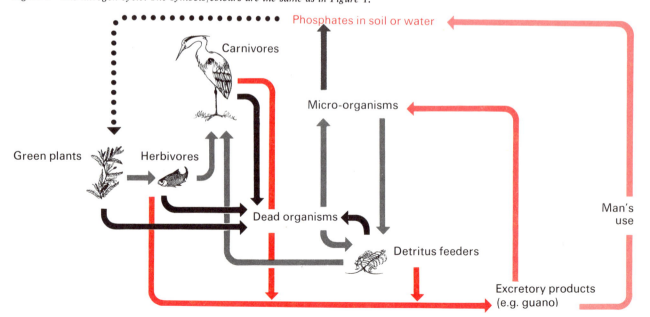

*Figure 4* The phosphorus cycle. The symbols/colours are the same as in Figure 1.

## 20.2 Niches and Communities

Studies of production ecology, of the flow of energy and of the cycles of carbon, nitrogen and phosphorus show that there is a definite structure to communities based on food webs. There is a relationship between the plants (primary producers), herbivores, carnivores, detritus feeders (all consumers) and the micro-organisms causing decay; this relationship is expressed by the pyramids of energy, biomass and numbers. Each of the trophic levels just listed usually includes many species, and the lists of species vary between places and may change with time in any one place. So let us examine the composition of communities in greater detail.

One important attribute of an organism is its size. Consider a carnivore: this organism preys on other organisms.

**What relationship in size do you expect between a predator and its prey?**

In general, the prey will be smaller than, but of the same order of size as, the predator—as a mouse to a cat, or a snail to a thrush, or a lamb to an eagle. If the prey is much larger, then the predator will not be strong enough to kill it and if the prey is very small, then the predator will not be able to catch sufficient numbers to satisfy its needs.

There are exceptions: by hunting in packs, wolves or army ants can prey on relatively much larger animals, and, by having a special food-collecting device, blue whales can feed on 'krill' about 2.5 cm long.

Thus the size and structure of a carnivore sets limits to its prey species; so does the behaviour of the carnivore—a slow-moving starfish can prey on coral animals which do not move at all but it cannot prey on small fish, whereas a spider, motionless on its web, can catch an active fly. Thus carnivores are specialized and consequently limited in their choice of food.

**Do you expect herbivores also to be specialists and limited in their choice of food? If so, suggest examples.**

Certainly herbivores are specialists. Sheep and cows graze on grass; snails, locusts and caterpillars also eat plant leaves; aphids (greenfly and blackfly) suck plant juices; finches and other birds eat seeds; squirrels and pigs eat acorns. These different habits require different structures and different types of behaviour.

**Can you state a general relationship between the sizes of herbivores and their food plants?**

There is a lower limit of size below which the herbivore will not be able to satisfy its needs (unless it has some unusual specialization, as for example for feeding on phytoplankton), but there is no obvious upper limit of size.

Detritus feeders are still less limited in the size range of their food than are herbivores. The dead bodies of whales and elephants are disposed of equally with those of crabs and mice. Parasites, on the other hand, are limited to hosts that are larger than themselves, usually much larger.

## 20.2.1 Niches

Specialized habits and activities of organisms are the basis for the concept of 'niche', proposed by Elton in his classic book *Animal Ecology* in 1927. In a human community such as a town or village, there are recognizable niches such as policeman and publican and shop-keepers of different kinds. Each title conveys the function of the individual and some of his attributes, e.g. the policeman must be over a minimum height, but the others could be tall or short. Similar niches may be present in human communities in different parts of the world, showing interesting similarities and differences —for instance, monarchs, presidents and prime ministers. For animals, the niche can often be defined by its size and food habits, but other habits are also relevant.

niches

When communities of organisms have been broken down into niches, the species filling these in different places can be identified and compared. Take, for instance, the herbivores feeding on oak-trees: these can be classified into large herbivores and small herbivores, both eating leaves; herbivores feeding on sap; herbivores living as parasites in the leaves; herbivores feeding on other plants which grow on the oak such as the alga *Pleurococcus* (which forms a thin powdery green covering over the bark of the tree usually on the north-facing side) and lichens.

List these herbivores if you can from general knowledge by filling the spaces below:

(a) large leaf-eaters:

(b) small leaf-eaters:

(c) small sap-eaters:

(d) parasites:

(a) deer
(b) caterpillars
(c) aphids
(d) gall insects

The herbivores on *Pleurococcus* and lichens are woodlice, caterpillars, millipedes and slugs.

From your general knowledge, identify similar niches and the animals filling them on cultivated herbs and grasses (i.e. on fields of crops or market gardens or parks and family gardens).

large leaf-eaters: cattle, sheep; sometimes deer, rabbits.
small leaf-eaters: caterpillars, snails, slugs, beetles, grasshoppers.
parasites: eel-worms, root-fly larvae.
small sap-feeders: aphids, scale insects.

The smaller plants are probably also present but growing on the ground, not on the cultivated plants—they will have similar herbivores to those in the wood. You will have realized that the second list is very similar to the first list—it is only when the *species* are listed by name that the two lists look very different. For instance, compare these lists of common caterpillars found on various plants:

| *oak* | *pine* | *cabbage* | *nettles* |
|---|---|---|---|
| winter moth | bordered white | cabbage white | small tortoiseshell |
| green tortrix | pine beauty | diamond back moth | pearl moth |

All eat the leaves and have similar jaws; they all have similar life-histories —the full-grown caterpillar forms a pupa (chrysalis) and later emerges as the adult moth or butterfly; the adult lays eggs which hatch into caterpillars. So the feeding niches are identical, but filled by different species on the different plants.

caterpillars and plants

Some of the insects which live on oak-trees will live only on this species, whereas some are able to eat the leaves of several or many species of tree. Each tree species has associated with it some insect species which feed only on it.

The specificity of herbivores is a very important facet of community structure. It explains why some animals are absent from some places—simply because their food plant is absent.

All the examples of niches given so far have been of herbivores, but there are plenty of other types of niche in the other trophic levels, as well as further herbivore niches not listed here. Elton quoted some examples showing remarkable parallels in very different places. One is the 'tick-picker'—a carnivore that picks parasites off the outside of their hosts. In Africa, the tick-bird feeds entirely on the ticks that live on the skin of large herbivores such as elephant, rhinos and antelopes. In England, starlings often pick parasites off sheep and deer. In the Galapagos Islands (which you will learn more about in Unit 21) a scarlet land-crab picks ticks off the skin of the marine lizards. In the lagoons of coral reefs, small striped fishes pick parasites off the skin of large fish of many species; the latter queue for the attention of the 'cleaner fishes'; these live in definite places and have a very characteristic darting movement. In some other reefs, shrimps perform the same function of picking parasites off fishes. The existence of this curious niche clearly implies specialized behaviour on the part of the tick-picker and of the larger animal being cleansed of its parasites. The myriad other niches also imply specializations of structure or behaviour or both; sometimes, as with the tick-pickers, they are very 'narrow', but sometimes they are wide.

tick-pickers

## 20.2.2    The distribution of plants

Since carnivores prey on herbivores, often in a specialized way and hence on certain species of herbivore only, and herbivores eat plants, but are often highly specific and restricted in their diet to a few, or only one, species of plant, the distribution of plant species governs the distribution of animal species. The basic nutritional requirements of plants are very simple—carbon dioxide, water and mineral salts—so their distribution might be expected to be either uniform or random. But travel through different parts of the world or look at pictures of them and you will realize that the wild vegetation is different in different places. The crops also are different. Bananas grow only near the tropics, whereas apples and pears grow freely in Britain, Canada, Tasmania and New Zealand. Oranges and other citrus fruits are produced in the Mediterranean countries as well as in California and South Africa.

**What are the differences between these various countries which might explain their differences in vegetation?**

Differences in 'climate' sum up a number of differences in factors affecting plant growth and survival.

Three important factors are rainfall, daylight and temperature. Some areas have no rain for years—Arica (Chile) on the west coast of South America has an annual average rainfall of 1 mm—whereas more than 3 000 mm falls annually on others. Some places have very definite seasons —Freetown (Sierra Leone) on the west coast of Africa has an annual rainfall of 3 990 mm and more than half of this falls in the three months July, August and September—whereas in others there is rain evenly dispersed through the year. Compare these extremes with London (annual average 620 mm) and Seathwaite, Cumberland (annual average 3 290 mm, mostly falling in the winter).

rainfall

22

Light is essential for photosynthesis. You are aware of the change in day-length and in the intensity of sunlight in the British Isles during the year.

**light**

**How do the day-length and intensity of sunlight vary**
**(a) close to the Equator, (b) in 'high' latitudes, more than 67° north or south of the Equator?**

(a) There is very little change in either day-length (about 12 hours) or the maximum intensity of sunlight (but there may be seasons with many cloudy days).

(b) There is a contrast between midsummer with 24 hours of daylight and midwinter with continuous night; the intensity of sunlight varies from nothing to a high level.

**What differences in plant growth would you expect to find between places with equal day-length through the year and those with day-lengths varying between 0 and 24 hours?**

You must be aware that most plants in the British Isles grow between about March and October—for many of them, the amount of light is insufficient for growth in the winter period. Plants growing in higher latitudes have shorter growing periods.

When there is no light, plants must cease to grow, so plants in places with variable day-length would be likely to grow very well during the summer part of the year and stop growing completely in winter; in contrast, with no variation in day-length, the plants should grow uniformly through the year if other factors allow this.

Of course, the changes in amount of light in high latitudes are accompanied by changes in temperature. Plants living there are subjected to long periods of freezing cold in winter, and the adverse effects of this must be added to the adverse effects of the limited growing period.

**temperature**

In tropical regions, great seasonal changes in rainfall may control seasonal growth of plants. Some tropical plants, however, are very sensitive to changes in day-length and respond to differences as little as five to fifteen minutes.

Thus from a study of rainfall, day-length and temperature through the year, it should be possible to predict how favourable for plant growth the conditions are in different parts of the world. Look at Figure 5 which shows the natural vegetation typical of different areas of the world (p. 24).

In the very high latitudes of the northern hemisphere is the treeless tundra, dominated by lichens. South of this is a belt of evergreen coniferous forest, and south of this are forests of broad-leaved deciduous trees (i.e. trees which lose their leaves in winter). In those parts of the tropics where there is a high rainfall, there are luxuriant forests. Between these forests and the temperate deciduous forests lie the grasslands called savannahs and steppes, and evergreen woodlands. Where the rainfall is very low or where it is very erratic and varies from year to year, there are the great deserts of the world. These vegetational zones are named after the *climax* vegetation— the natural vegetation which should grow if there were no human or other interference. Thus, in Britain, almost any area will revert to woodland if left uncultivated. Even a lake will gradually fill up with silt and turn into a marsh and then a fen and finally into woodland. In upland Scotland, the wood will probably be coniferous, but English woodland will be of broad-leaved trees.

**vegetational zones of the world**

Within the broad vegetational zones defined by climate, there are variations in the distribution of wild plant species. This can be seen even in the

23

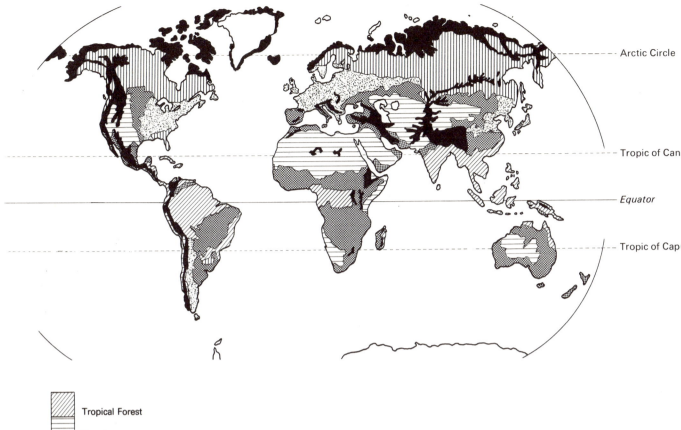

Tropical Forest

Desert

Savannah, Steppe, Non-coniferous Evergreen Forest

Deciduous Forest

Coniferous Forest

Tundra and Alpine

*Figure 5   Map to show the main vegetational zones of the world.*

small area of southern England. Tansley, in his monumental work *The British Islands and their Vegetation*, describes the oak-woods as consisting typically of:

oak-woods

1  a tree layer—mainly oaks, with ash and sycamore frequently present;
2  a shrub layer—mainly hazel, with hawthorn, blackthorn and others frequently present;
3  a field layer—the botanists' 'herbs'. There are many species which may be present but not all in the same wood. Contrast these two lists of common herbs:

| List A | List B |
|---|---|
| *Pteridium aquilinum* (bracken) | *Mercurialis perennis* (dog's mercury) |
| *Rubus 'fruticosus'* (blackberry) | *Urtica dioica* (stinging nettle) |
| *Endymion non-scriptus* (bluebell) | *Circaea lutetiana* (enchanter's nightshade) |
| *Agrostis tenuis* (grass) | *Fragaria vesca* (wild strawberry) |
| *Holcus mollis* (grass) | *Brachypodium sylvaticum* (grass) |
| | *Deschampsia caespitosa* (grass) |

24

**Suggest an explanation for the difference between them.**

If you are a gardener, you probably realized that these herbs come from woods growing on different types of soil.

Soil composition, texture and drainage are included in *edaphic* factors (from the Greek for 'soil'). List A is from oak-woods growing on an acid, sandy, dry soil where the shrub layer is often rather scanty; list B is from damp oak-woods growing on alkaline, clay soil where the shrub layer is usually well developed.

edaphic factors

Look at film strip 20(b): 1, 2, 3 and 4. These illustrate the effect of the spacing of the trees on the field layer (the herbs) of a woodland.

The distribution of plants also depends on their past history and on the other species present. Each species evolved in a limited geographical area and its spread from this area depends on some mechanism for dispersal; this is often either by wind-blown seeds or by animals eating fruits. Man, the cultivator and traveller, has introduced many species into areas which they had not colonized unaided. A plant arriving in an otherwise suitable locality may fail to establish itself, either because it cannot compete with the plants already growing there or because it is eaten out by herbivores. Similarly, animal species arriving in a new locality may fail to survive, either through lack of suitable food or shelter, or because of the presence of competitors or predators. Those organisms that do establish themselves have succeeded in finding and occupying suitable vacant niches or in replacing resident species.

dispersal

To sum up:

The presence or absence of a species of animal or plant depends on many factors:

(i) whether the species reaches that particular place;

(ii) whether climatic and edaphic factors are suitable;

(iii) whether the appropriate food is present;

(iv) whether the species can survive in the face of competition or predation.

*Now you can do* SAQs 5 to 7.

## 20.3 Numbers of Individuals in Populations

One of the vital activities of organisms is reproduction. The survival of a species depends on successful reproduction of individuals. Since the survival of successfully adapted species is the basis for evolution, successful reproduction of individuals is also the basis for evolutionary change. Some organisms reproduce by simple division or by the formation of buds. Only one individual need survive long enough to reproduce for the species to be perpetuated to the next generation. You should remember (Units 17 and 19) that this type of reproduction implies that all the descendants will resemble their ancestors genetically; there is no inbuilt mechanism to produce diversity. Many flowering plants propagate themselves in this way, but it is very uncommon among animals.

The great majority of animals and most plants reproduce sexually; in Unit 19 we discussed this process at the chromosomal level and showed its importance from the point of view of genetics. In most of this discussion of population numbers, and in the following section on population dynamics, we shall generally be concerned with organisms which reproduce sexually, and so with life-histories which start with a zygote (fertilized egg) and, if successful, end with the new adult playing its part in sexual reproduction, either by laying eggs or by fertilizing eggs produced by another individual.

**Table 1**

**Average number of fertilized eggs produced by each female**

| | |
|---|---|
| oyster | $100 \times 10$ |
| codfish | $9 \times 10$ |
| plaice | $35 \times 10$ |
| salmon | $10 \times 10$ |
| stickleback | $5 \times 10$ |
| winter moth | 200 |
| mouse | 50 |
| dogfish | 20 |
| penguin | 8 |
| elephant | 5 |
| Victorian Englishwoman | 10 |

Suppose that there is a population of five pairs of organisms (five females and five males)—if this population is to produce an offspring generation of exactly the same number of adults, what is the mean number of eggs that each female must produce?

But you learnt in Unit 19 that there is generally an over-production of individuals in each generation. Look at Table 1 which shows the average number of fertilized eggs produced in their lifetime by females of different organisms (this is called *fecundity*). Answer the following question:

Assuming that every egg is fertilized and that every egg grows into an adult, then the number is two. In producing two eggs each of which grows into an adult, the female is providing a replacement for herself and a replacement for one male.

For each population to remain stable (i.e. for the number of adults in the off-spring generation to remain the same as in the parent one), how many eggs from each female (on average) must survive?

Two.

For each species, write down the number of eggs which must die before becoming adult, if the population is to remain stable. Then express this number as a percentage of the total number produced.

e.g. penguin—6 must die if only 2 ca[n] survive: $6/8 = 75$ per cent.

The last set of numbers you have written down are called the *pre-reproductive mortality* for each generation. The word *mortality* is used to measure the 'rate of death'—the percentage dying (in this case between egg production and reproduction of adults of the next generation). Subtract this figure from 100, to measure *survival* for the same period. Thus, for penguins the survival to the reproductive stage is $100 - 75 = 25$ per cent; for mice it is $100 - 96 = 4$ per cent. The fecundity, when expressed

as a proportion of the adult population (this is called birth rate when applied to mammals), measures the inherent ability of the population to increase. The pre-reproductive mortality measures the rate of death of a generation.

**What is the relation between fecundity and pre-reproductive deaths if the population is stable (numbers of the offspring generation able to reproduce = numbers of parent generation)?**

Fecundity minus deaths should equal the number in the parent generation.

If (births minus deaths) > than parent numbers, there will be more individuals in the offspring generation than in the parent generation. If this is repeated for several generations, the population will keep increasing exponentially.

Conversely, if (births minus deaths) < parent numbers and this state of affairs persists for several generations, then that population will keep on decreasing.

*If you need to convince yourself of this, turn to* SAQ 9.

Usually the death rate is more variable than the birth rate. Thus, changes in the numbers of breeding adults are generally a result of changes in mortality; so, in most of what follows, we shall be concerned with factors causing mortality.

birth rate and death rate

In the discussion so far, we have taken as the unit of time 'a generation'—that is, the interval between the production of eggs by parents and offspring. Table 2 shows that this varies greatly with species and may vary with external factors such as temperature. In practice, times are generally measured as recorded by clocks and calendars, but it is very important to interpret these from the biological point of view in relation to the duration of each generation.

**Table 2**

**Average generation times for various species**

| *Species* | *Generation times (averages)* |
|---|---|
| man | 20 years |
| elephant | 20 years |
| royal albatross | 10 years |
| codfish | 6 years |
| salmon | 4 or 5 years |
| great tit | |
| stickleback | 1 year |
| winter moth | |
| mouse | 8 weeks |
| *Drosophila* (fruit fly) | 3–4 weeks |

We have rather assumed so far that generations are discrete—that is, the lives of the parents and offspring scarcely overlap in time. This is the rule for thousands of species of British insects that have an annual life cycle. Take as example the winter moth: this gains its name because the adults are active in November and December. The males have normal wings; they fertilize the eggs as the wingless females climb up the trunks of oak-trees by night. The eggs are laid on the oak twigs and the adults die; the eggs hatch in spring, in March or April, into green caterpillars; these feed on the oak leaves and when full grown, usually in May, let themselves down to the ground on silken threads. They burrow into the ground and each forms a chrysalis or pupa from which the adult emerges on a 'warm' night in the following November. Since the adults die after the eggs are laid, there is no overlap between the active adult and the next active stage, the caterpillar; each generation takes one year.

*winter moth life history*

Look at film strip 20(c) for photographs of winter moth adults, caterpillar and pupa. The life cycle is illustrated in Figure 8 (p. 33).

For many other species of insects, and for most fishes, birds and mammals, the generations overlap. Adults may produce several separate broods of offspring and their older offspring may breed while the parents are still breeding. Take as example the domestic cat.

**If a female cat produces her first litter of kittens when she is exactly one year old and her offspring behave in exactly the same way, how many generations of her offspring could be alive in the month before she reaches her seventh birthday?**

An analysis of a cat population with overlapping adult generations and a long adult life is clearly more complicated than an analysis of a winter moth population with discrete generations and an annual life cycle. This explains why most of the examples we shall use in discussing population dynamics are studies of insects (these are also important because they are 'pests').

| Mother cat's birthday | Offspring new born | Generations adult |
|---|---|---|
| 0 | — | — |
| 1 | first | — |
| 2 | second | first |
| 3 | third | first and seco |
| and so on: | | |
| 6 | sixth | first to fifth |

So the answer is that six generations of offspring could be alive just befor the mother cat's seventh birthday (ar she could well still be breeding herse and producing three litters a year!).

Pre-reproductive mortality covers all the deaths from whatever causes between the laying of a batch of fertilized eggs and the maturation of the surviving organisms which emerge from them. Some eggs may die before hatching and individuals may die at all ages between hatching and maturity. Thus there will be a decrease in the number of survivors with time; when plotted on a graph, this is called a *survivorship curve*.

**Which of the three descriptions refers to each of the survivorship curves of Figure 6?**
**(a) There is high mortality soon after the eggs hatch and thereafter the numbers fall off gradually; some adults live for several years.**
**(b) There is high mortality soon after eggs hatch; the adults live for a few days only.**
**(c) The numbers fall off gradually until near the end of the life span, when the adults begin to die in large numbers.**

(a) describes the curve for the trout;
(b) describes the curve for the winte moth;
(c) describes the curve for man.

Many species of animals suffer a high mortality early in life (like the trout). When there are marked changes in body form, as in winter moth (or when tadpoles change into frogs), there are often also periods of high mortality. Survivorship curves form the basis for tables of 'life expectancy' such as those used by actuaries computing rates of human life assurance.* These

* See Table 2 (p. 10) of Population Dynamics, by M. E. Solomon, for a human life table (for the U.K.). Figure 3.1 on the same page shows three survivorship curves.

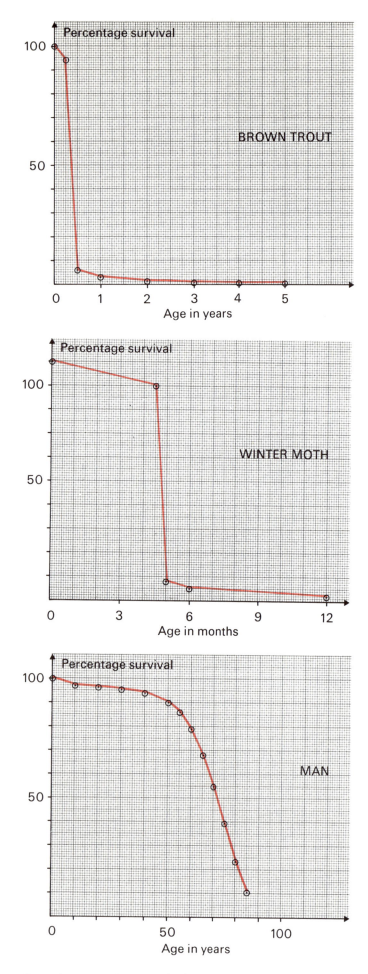

Figure 6 *Survivorship curves for three organisms.*

tables give the mean expectancy of life for individuals of each age. Constructing such life tables is an essential step in the investigation of population dynamics (see later). They make it possible to deduce 'age-specific mortalities'—that is, the rates of death for groups of individuals of different ages (or at different stages of the life cycle). Knowing when individuals are most likely to die makes it possible to identify the important causes of death and thus the factors affecting the size of populations.

### 20.3.1 Population curves

First let us consider what is likely to happen when a species colonizes a new and initially favourable environment. When a bacterial culture is inoculated* on to an appropriate culture medium, the bacterial cells multiply through each cell dividing into two daughter cells (note that this is asexual reproduction).

---

**Assuming that 50 cells are introduced into a culture medium and that each cell proceeds to divide into TWO daughter cells every 20 minutes, calculate how many cells are present after: 1 hour, 3 hours, 24 hours.**

There are 3 divisions every hour. After 1 hour, the 50 cells will have increased to $50 \times 2^3 = 400$ cells. After 3 hours, the number will be $50 \times 2^9 = 25\,600$ cells. After 24 hours, number will be $50 \times 2^{72} = 2.35 \times 10^{22}$.

---

**You have met a term which describes this type of regular increase in numbers—name it.**

Exponential increase—refer back to Unit 2 and see *HED*.

---

**If you were to plot the numbers of bacteria for the first three hours on an arithmetic scale and on a logarithmic scale, would either of these plots give a straight-line relationship between numbers and time?**

Yes, the plot on a logarithmic scale would give a straight line.

---

Look at *Population Dynamics*, p. 12, Figure 3.2. Data for the exponential increase in numbers of a small mite that lives in grain are plotted on the two types of scale. Figure 3.2 (b) is almost a straight line—the numbers are plotted on a logarithmic scale.

Since the bacterial culture must be in a vessel of limited size, and the amount of medium present is limited, there must come a stage when the numbers present cannot be doubled every twenty minutes—the individuals are 'competing' for the limited 'resources' of space and medium. The rate of increase must fall off. More than 100 years ago Verhulst derived an equation to describe the rate of growth of a population in a limited environment, assuming that the resources can support a certain maximum density. When plotted on an arithmetical scale, the numbers of individuals in such a population give a curve called the 'logistic curve'—look at Figure 3.3 (p. 15) of *Population Dynamics* for an example.

**logistic curve**

About 50 years ago, Pearl showed that numbers of yeast in a sugar solution followed the logistic curve. The same type of increase occurs when algal cells (such as diatoms) are added to an appropriate culture medium; it probably also occurs in temperate and arctic lakes and oceans when phytoplankton begins to 'grow' in spring. All these organisms reproduce by each cell dividing into two. When organisms which reproduce sexually

---

* *Transferred from a stock culture, e.g. on a loop of wire or in a sterilized pipette; both methods were shown in TV 17.*

and produce many young at a time enter a new environment, their numbers may increase in a way that resembles the logistic curve, but usually the values do not stay constant after reaching the highest value. Typically there are fluctuations, often about a slightly lower level. One example of 'natural' conditions under which organisms may colonize new environments is when pest insects, such as flour beetles and moths, reach stored products; there have been many observations and experiments on these pests, particularly flour beetles.

For an example of a natural community, look at Figure 7 for population changes of caterpillars on oak-trees in Wytham Wood near Oxford. All these species have annual life cycles with discrete generations, and all feed at the same time of year on oak leaves. Note that the numbers are plotted on a log scale. Answer the following questions:

**What is the proportion between the lowest and highest numbers for: winter moth; green tortix; November moth? Do the numbers of the caterpillars of different species vary in the same way or in different ways?**

So, when we discuss population dynamics in section 20.4, we shall try to explain why some organisms are common and others are rare, as well as trying to explain why the numbers of individuals of all species fluctuate from year to year (generation to generation).

All the values fluctuate in roughly the same way—there are years when all the caterpillars are numerous, and years when all are few; the ratio between maximum and minimum numbers recorded is roughly 100 to 1 for all the species, but some species are always more numerous (i.e. these are common) whereas others are always less numerous (these are rare).

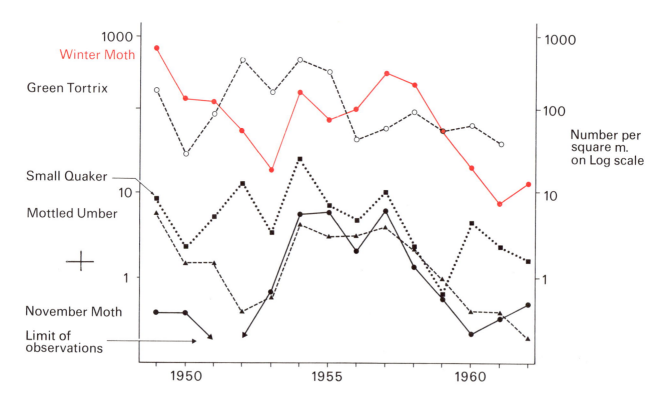

Figure 7  *Population changes for some caterpillars on oak-trees in Wytham Wood, near Oxford. The numbers are plotted on a log scale. See film strip 20(c): 3, 6, and 7 for photographs of three of these caterpillars.*

## 3.2  Sampling populations

Laboratory experiments are set up so that all the individual organisms can be counted and perhaps measured at various stages. Thus, accurate figures

are obtained for the changes in number and size and shape of individuals during the course of the experiments. Even so, in replicates (repeats) of the experiment, using organisms of the same species, same population, even of the same family, the figures are very unlikely to be identical. In your first home experiment, you found that physical measurements are not exactly accurate. Here we are dealing with biological material of great complexity (whole organisms) so it is not surprising to find some variability. There are now many statistical tests available for assessing whether the differences between observations are significant or could have arisen by chance.

The same problem of inherent variability must arise in field investigations comparing, say, the caterpillar population of two adjacent oak-trees or in two separate woodlands. A further difficulty in field studies is the improbability of collecting and examining every single organism in the population. Indeed, such a procedure might disrupt completely the natural sequence of events. So estimates of natural populations are based on *samples* (like public opinion polls).

population estimates

There are many published papers on methods of collecting organisms and making such collections quantitative; there are statistical tests and computer programmes for deducing from a sample the probable size of the population from which it came, and for obtaining the probability of this being a reliable estimate. The methods used must be related to the habitat being studied and the life-history and behaviour of the organisms under observation. Consider first a plant community growing on the chalk downs:

---

**How would you set about estimating the numbers of a given plant here?**

---

Rooted plants seem easy to count—but walking through the tangled undergrowth of a forest is very different from marking squares on chalk grassland, and there can be practical difficulties in carrying out a census of plants. With animals which move about, perhaps avoiding the investigator, the difficulties are obvious.

You could crawl all over the downs, counting the individual plants— perhaps marking each one on a map. But probably you would sample certain areas—perhaps squares of one metre sides—counting the number in each of several squares and then applying the appropriate statistical method to get probable estimates for the whole area.

The life-history of the winter moth is shown in Figure 8. There are two periods in its life when individuals move in definite directions and this makes it possible to census the population twice in each generation. The females crawl up tree-trunks and can be caught in traps like inverted lobster pots which each catch insects crawling up one eighth of the circumference of the tree. The full-grown caterpillars drop down on silken threads from the branches to the ground and can be caught in trays of definite area under the trees. For other stages, the counts are less accurate—eggs can be counted on twigs, but it is difficult to estimate how much of the area used by females for laying eggs is represented by the collected twigs.

Full-grown caterpillars can be dissected to give the percentages diseased or attacked by parasites. One parasite attacks the pupae in the ground; adults of this parasite and of others can be collected by inverting a metal tray of known area with glass tubes fixed in its corners. The adult parasites go towards the light and therefore collect in the tubes. The numbers collected give a value for density of parasites related to the ground area; the percentage of pupae attacked can then be derived from the known numbers of larvae which descended over that area of ground. These population estimates all assume that the sample areas are typical of the whole area and it is essential to calculate limits of probability using statistical methods.

winter moth census

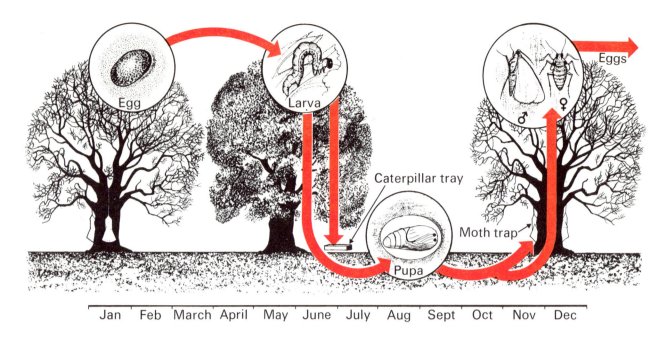

| Jan | Feb | March | April | May | June | July | Aug | Sept | Oct | Nov | Dec |

*Figure 8   The life cycle of winter moth and methods of sampling its population. See film strip* 20(c): 2 *and* 3 *for photographs of live winter moth adults and caterpillar; and* 20(c): 4 *and* 5 *for parasites of winter moth and empty pupal cases.*

For each species studied, special methods of trapping must be devised to be used in appropriate places and at appropriate times. Small mammals such as voles and fieldmice are caught in box traps; birds may be caught in traps or in mist nets or persuaded to use nest-boxes; fish may be trapped or netted or caught by using electrical gadgets. The animals may be examined alive and then released or may be killed for detailed examination. The recapture of animals which have been marked and released forms the basis for some methods of estimating populations; the subsequent examination of animals caught on previous occasions gives positive information about growth and survival.

traps

Populations of animals with long lives and overlapping generations present special problems since it is desirable to break down the total population numbers into those for different age or generation groups. So it is necessary to deduce the ages of the animals in the samples and to assume that these are typical of the population as a whole.

You can read *Population Dynamics*, chapters 1, 2 and 3 as a black-page Appendix now if you like.

*Now you can do* SAQs 8 *to* 11.

## 20.4 Population Changes

We have mentioned two sorts of change in number of individuals in a population:

1 Within each generation, there is a reduction in numbers from the eggs to the breeding adults.

2 There are fluctuations in numbers from generation to generation; these can be revealed by counts of the same stage of the life-history in successive generations—i.e. counts of caterpillars over a number of years, or of moths, or of kittens, or of five-year-old cats.

Look at p. 17 of *Population Dynamics*. Figure 4-1 illustrates both these types of changes, using data for a grasshopper population. Figure 4-2 shows the second type of fluctuation in numbers, based on counts of the pupae of a moth; note that the full line is for the actual counts of pupae per 100 m², and the dashed line shows the same numbers plotted on a logarithmic scale.

Fluctuations in numbers from generation to generation could depend only on breeding success of the adults—on the number of eggs that the average female is able to produce. This situation may operate for species of birds with special nesting sites. If all available sites are occupied, then no further females are able to lay and incubate eggs; thus the number of nesting sites sets a maximum limit to the number of eggs. If the generation suffers little mortality, then there will be an excess of birds over sites. The number of nests will be the maximum possible and some birds will be unable to breed. If there is heavy mortality within the generation, then there may be unoccupied nesting sites, and egg production will be less than the maximum possible.

Many birds show territorial behaviour during the nesting season. Each pair, or sometimes only the male bird, reacts in a characteristic 'threatening' way to any other bird of the same species which enters the area round the nest; the intruders usually retreat at once. The threat may be expressed by a gesture, such as the display of the robin's red breast, or by song, or in both ways. Sometimes the territory is just a nesting area and **territory** it may be very small, as is that of gulls in their crowded breeding colonies; but often the territory is also the area in which the parent birds find the food for themselves and their young. The system of territories spaces out the breeding pairs, setting a maximum limit for any area. Usually the territories vary in size; they are smaller when the potential breeding population is large and larger when this population is small. Other animals besides birds also display territorial behaviour, for instance, such fish as the male stickleback.

The degree of crowding of the adults may also affect reproductive success. When voles* are crowded together, the females produce on average fewer **crowding** young than when they are not so crowded. This is an example of an effect which is 'density dependent'. If the average number of young produced are plotted against the degree of crowding of the adults, there is an inverse correlation between the two. See Figure 4-8 (p. 21) of *Population Dynamics* for similar figures for a great tit population. For the voles and the tits,

* *Voles are herbivores resembling mice.*

there is 'negative feedback' between the numbers of adults and those of young produced: an increase in the adult population means that fewer young are produced by each female.

### What will happen if the population of adults decreases?

So far the discussion has assumed that pre-reproductive mortality remains the same whatever the size of the population; but, as stated earlier, most investigations have revealed that changes in fecundity (or birth rate) are less dramatic and probably less important in population fluctuations than changes in death rate (mortality).

The average number of young produced per female will increase and the population will rise again.

## 4.1 Key factors

Haldane showed that calculations involving mortality and survival can be simplified by using logarithms. He represented the killing power of a mortality factor by the change in the population density caused by it measured on a log scale. This logarithmic measure of killing power is termed the $k$-value; it expresses the reduction in numbers of individuals, per unit area, caused by one mortality factor.

**$k$-value**

The total pre-reproductive mortality, $K$, is the sum of the killing powers of the mortality factors acting in succession on the population of eggs produced by the parent generation. Thus,

$$K = k_1 + k_2 + k_3 \ldots k_n.$$

Each $k$-value is the difference between the logarithms of the numbers of individuals per unit area before and after its action.

Take the winter moth as an example. Refer to section 20.3.2 if you need to remind yourself about its life-history (shown in Figure 8, p. 33) or the methods used to sample its numbers and those of its parasites. Suppose that the population is not changing, so that the number of breeding adults per unit area is the same for the parent and offspring generations. Let us consider the offspring of a single female as occupying unit area.

### Then, if each female produces 200 eggs, what is the value of $K$?

Of these 200 eggs, only 2 can survive so:

$K = \log 200 - \log 2 =$
$\qquad 2.3010 - 0.3010 = 2.0$

or, $K = \log \dfrac{200}{2} = \log 100 = 2.0$

Here is a table for the winter moth, showing the average number of individuals killed by six mortality factors acting in succession:

| | |
|---|---|
| Number of eggs laid by female moth | 200 |
| Number killed by 'winter disappearance'* | 184 |
| Number of caterpillars killed by a parasitic fly | 1 |
| Number of caterpillars killed by other parasites | 1.5 |
| Number of caterpillars dying from disease | 2.5 |
| Number of pupae killed by predators in the soil** | 8.5 |
| Number of pupae killed by a parasitic wasp | 0.5 |
| Number of adults surviving to breed | 2 |

Film strip 20(c) includes photographs of the fly (number 4) and the wasp (number 5).

* 'Winter disappearance' includes the eggs dying from all causes (probably a small number), and a heavy mortality of young caterpillars directly after they hatch from the eggs.

** The pupae in the soil are eaten by various beetles and by shrews.

From these figures for winter moth, the following $k$-values are derived:

$k_1$ = the killing power of 'winter disappearance'             = 1.10

$k_2$ = the killing power of the parasitic fly that attacks the caterpillars = 0.03

$k_3$ = the killing power of other parasites that attack the caterpillars = 0.04

$k_4$ = the killing power of the caterpillar disease           = 0.09

$k_5$ = the killing power of predators that attack the pupae      = 0.64

$k_6$ = the killing power of the parasitic wasp that attacks the pupae = 0.10

Total pre-reproductive mortality $= K = k_1 + k_2 + k_3 + k_4 + k_5 + k_6$    = 2.00

Since we have used average values, $K = 2.00$ and the offspring generation of adults would equal in numbers the parent generation; the population would remain constant. Look back to Figure 7. The numbers of caterpillars of winter moth fluctuate greatly from year to year; this suggests that $K$ fluctuates. If this is so, then at least one $k$-value must fluctuate.

Figure 9 shows how $K$ and the successive $k$-values varied between 1950 and 1961.

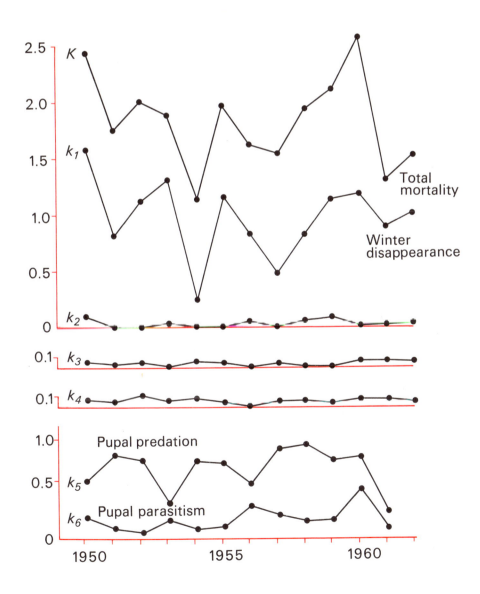

*Figure 9*   K *and* k-*values for winter moth in Wytham Wood between* 1950 *and* 1961.

It is obvious that the values for $k_1$ (winter disappearance) fluctuate in almost exactly the same way as the values for $K$ (total mortality). This means that changes in $k_1$ account for almost all the change in $K$ from year to year so $k_1$ (winter disappearance) is called the *key factor*. It is the major cause of the enormous fluctuations in the numbers of winter moth caterpillars. Although part of this mortality may be due to death of eggs, most of it occurs because the very young caterpillars do not establish themselves and start to feed. The eggs hatch in spring at about the time that the buds of the oak-tree open and its young leaves begin to grow. If the caterpillars hatch before the buds burst, they starve. As a result of natural selection, the hatching time of winter moth eggs is almost synchronized with the bud burst of oak-trees; but slight differences in timing lead in some years to very successful establishment of caterpillars and consequent defoliation of the oak-trees and in other years to high mortality of young caterpillars. You may wonder why the hatching time is not delayed—probably, if the caterpillars hatched too long after the buds burst, there would be a mortality later, because the leaves become tough and loaded with tannin, and many caterpillars would fail to reach the stage of being ready to pupate. So there would be selection against later hatching.

**caterpillar survival**

## 0.4.2 Regulating factors

Winter moth numbers fluctuate about a mean level. Figure 9 shows that changes in $k_1$ are usually greater than changes in $K$: this suggests that some other factor compensates by reducing the numbers more when $k_1$ is small and less when $k_1$ is large. This compensatory effect must be density dependent and it is acting as a *regulating factor*. There is negative feed-back, restoring the status quo.

**density dependent factors**

**Try to identify possible regulating factors from the graph of *k*-values for winter moth.**

There are two possible regulating factors: pupal predation and pupal parasitism. No other *k*-values show consistent relationship to *K*.

To make a convincing case for a certain mortality factor to be a regulating factor, it is necessary to show that its values vary in the appropriate way with the winter moth density to an extent that is statistically significant. This requires data from the same place for a number of years and it also requires a good understanding of statistical theory. For the winter moth, $k_5$ (pupal predation) is density dependent: when there are large numbers of pupae in the ground, then the predators prey on them more heavily. It appears that they concentrate on winter moth when it is abundant, but tend to ignore it when it is scarce. Thus the activity of the predators tends to stabilize the numbers of winter moth.

The pupal parasite is an ichneumonid* wasp. Many ichneumonids are parasites of insects. In this Unit's TV programme you will see one that is parasitic in flour-moth caterpillars. Most are highly specific, being able to live only on individuals of one host species. Although they are generally called parasites, they kill and usually consume the whole of the host, so that they are as much specific predators (each slowly consuming only one individual) as parasites in the usual sense of that word. The fly that parasitizes the winter moth caterpillar is also this type of parasite. The part played by such specific parasite-predators in population control is of special interest, because, if successful, they can regulate the numbers of their host species while having no direct effect on any other organism.

**insect parasites**

* *A special type of wasp with larvae that are parasites of other insects.*

There have been various theories and mathematical formulations to explain the interrelationships of insect host and insect parasite. Probably these parasites act as *delayed density-dependent factors*—that is, their reproductive rate depends on host density so that their own numbers are one or more generations out of phase with the host numbers. When numbers of hosts are high, the parasite is successful and the following year there will be large numbers of parasites. If the host is still plentiful, the percentage attacked by the parasite will be larger, because of the large number of parasites, so the host numbers will be reduced. The large generation of parasites emerging the following year will find fewer hosts to attack and so they will be less successful; the following year the parasite numbers will have fallen.

**delayed density-dependent factors**

In Figure 7, the numbers of different species of caterpillars can be seen to change from year to year, but some, such as the winter moth, are always common whereas others, such as the November moth, are always rare. The fluctuations from year to year are caused by variations in key factors. It is possible that all these species of caterpillar have similar key factors; their numbers fluctuate in roughly the same way. The difference in average population numbers of common and rare species depend on the regulating factors. It is obvious that regulating factors act so that November moth numbers never exceed 10 m$^{-2}$ whereas winter moth numbers vary about an average of about 100 m$^{-2}$. If winter moth numbers are sufficiently high for trees to be defoliated, it clearly becomes a 'pest'. Since this cannot happen if the regulating factors act to maintain small population numbers, it is of economic interest to be able to identify these factors. It may then be possible to manipulate them so that the insect becomes or remains rare: then it cannot become a pest. See 20.6.1 for a discussion of pest control.

**rare and common species**

### 20.4.3 Population models

The ultimate test as to whether key factors and regulating factors have been identified correctly is to simulate the inter-relationships by calculation, starting with the actual numbers counted in a certain year, inserting the various factors operating at the levels required by the current theory and then working out how the numbers should vary through the following years. The calculated figures can then be compared with those actually observed in the field.

This test has been applied to the winter moth at Wytham (see Figure 10). Starting with the observed value for winter moth caterpillars in 1950 and using each year the actual $k$-value for winter disappearance (since so far there is no method of predicting this from other observations), but representing the $k$-values for pupal predation by a density-dependent factor and using constant values for the other $k$ factors, calculations resulted in a set of figures which are not very far from those observed. Adding a $k$-value for parasitism, which varied as suggested by theory, improved the fit of the computer values in relation to the observed values. A new interpretation of the relation between parasite and host density and a consequent adjustment in the appropriate $k$-values has led to even better correspondence of observed and predicted values. From these new relationships, curves for the numbers of parasites to be expected were derived, and these vary in the same way as the observed values. This makes it very probable that the key factor and regulating factors for winter moth in Wytham Wood have really been identified and the latter accurately quantified.

The new interpretation of host-parasite relations was formulated from observations of the behaviour of an ichneumonid parasite called *Nemeritis*, whose host is the caterpillar of the flour moth. You will see flour moth caterpillars and adults of *Nemeritis* in this Unit's TV programme. In a

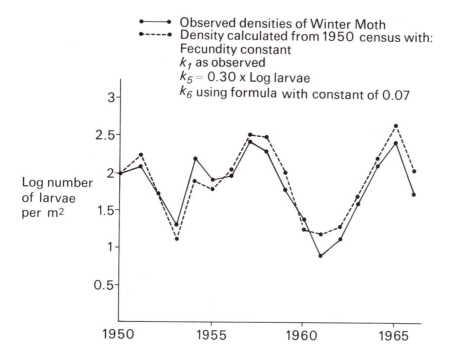

Figure 10  *Observed and calculated densities of winter moth in Wytham Wood. The calculated values were obtained by starting with the value for 1950, then: assuming constant fecundity; using the observed values for each year for $k_1$; taking $k_5$ as $0.30 \times log$ number of larvae; and calculating $k_6$ from a formula with a constant of 0.07.*

series of long-term observations on the relative numbers of host and parasite, it was found that the efficiency with which *Nemeritis* killed caterpillars decreased as the density of the parasite increased—the more numerous the parasites in a given area, the less effective was their infection of readily available hosts. Before these observations were analysed, it was assumed, because there was no reliable data, that the efficiency of parasites in killing their hosts would depend only on the host density and would not be affected by the density of the parasites.

Parasites such as *Nemeritis* seek out areas where there are many caterpillars rather than areas where caterpillars are few. They can discriminate between caterpillars that are already parasitized and those that are free of parasites or have been parasitized for less than ten minutes. They do not lay eggs in those already parasitized. These two types of behaviour should improve the efficiency with which *Nemeritis* attacks its host, the flour moth caterpillar. When many parasites gather in the same area, the later arrivals tend to fly away when they encounter parasites already there; the parasite that flies away may find another group of hosts, so the behaviour could improve the efficiency of the parasites' attack. If the parasites are very crowded, however, there may be no groups of hosts free from parasites and their response to mutual interference then means that the parasites' attack is less effective.

**host parasite relations**

Apart from its purely intellectual value in explaining observed data, a model should be able to predict what changes will occur when factors are manipulated.

Winter moth has become a pest of forest trees in eastern Canada. It had no parasites there, so one of the European parasites (the fly that attacks the caterpillar) was introduced. Winter moth numbers were reduced to low levels within five years. The population model developed at Oxford has been used to predict the likely further effects of this introduction—it predicts that there should be cyclic variations in numbers of winter moth and its parasite such that the trees should be defoliated (because of very

**winter moth in Canada**

large numbers of caterpillars) at intervals of nine or ten years. This would be an improvement on the earlier condition when the trees were defoliated almost every year.

There are various other records of insects in forests undergoing cyclic fluctuation and defoliating trees at intervals. Often the cycles are synchronized and defoliation occurs over a wide area. Because of their effects on the growth of valuable timber, there is great interest in the strategy necessary to limit the numbers of these pests.

Winter moth and the other defoliating caterpillars are all herbivores. In most years they are not limited by their food supply—the mathematical model simulating winter moth populations has no term measuring food supply. You may wonder if this is true of herbivores in general. Many of these are specific in their food plant and their distribution may be limited if that of the plant is limited. Probably few of the insects are limited in numbers by the abundance of their food supply during the main feeding period, but for many there may be a critical period when the young first begin to feed. For larger herbivores, the picture is not so clear. Grazing mammals certainly show preferences for certain plants, but will usually eat others if they are the only ones available. Some herbivorous mammals show regular cycles in numbers. The brown lemming* of North America builds up its populations every four years. The peak numbers are followed by a 'crash' to minimum numbers, and the populations then build up again to another peak. Investigations have shown that peak populations eat out their food supply and the crash allows the plants to recover. The European lemming has a similar four-year cycle in Scandinavia; this is well known because during the peak periods, swarms of lemmings migrate across country, swimming over rivers and often plunging to their deaths by drowning in fjords.

**lemmings**

### 20.4.4 Predator-prey relationships

The capture of the first meal is a critical stage for many predators; once the young animal has established itself where there is a source of food, its expectation of life is markedly increased. This has been shown, for example, for ladybird (beetle) larvae, which feed on aphids (greenfly) and for young brown trout, which eat shrimps and aquatic insect larvae. Some predators are specialists, as the ladybirds are on aphids, or mammalian ant-eaters are on ants and termites. Many predators, like the brown trout, have a fairly wide food range and can change their basic diet to take advantage of any suitable organisms that are abundant.

The maximum numbers of predators in an area must depend on the numbers of prey available. If the numbers of prey fluctuate, then it is very likely that the numbers of predators will also fluctuate. The changes in numbers of lemmings in Arctic areas, described in the last section, are associated with changes in the numbers of the Arctic fox which prey on lemmings; when lemmings are scarce, the foxes prey on ptarmigan and other birds, but the number of foxes falls. The snowy owl preys on European lemmings and is unable to rear young when the lemming population is low; its breeding success follows the population curve of the lemmings with their four-year cycle.

**Arctic fox**

**snowy owl**

Turn to Figure 4.3 (p. 18) of *Population Dynamics*. This shows the fluctuations in numbers of two mites (small organisms distantly related to spiders) in laboratory experiments. One mite, the prey, is herbivorous and is pro-

* *Small rodents related to voles and mice.*

vided with oranges as food; the other, the predator, feeds on the herbivorous mite. In this case, the increase in numbers of the predator causes the numbers of prey to 'crash' and consequently the numbers of the predator 'crash' soon afterwards. Reduction in the pressure of predation leads to increase in numbers of the prey and this is followed by increase in numbers of the predator; a 'crash' of prey numbers soon follows and the cycle continues to be repeated.

*Now you can attempt* SAQs 12 *to* 13.

If you wish, you can now read Appendix 1 (Black), which is a structured exercise based on a recent study of a tawny owl population near Oxford. The information is arranged to give you practice in applying some of the ideas discussed in sections 20.1, 20.3 and 20.4.

You may also read chapters 4, 5 and 6 of *Population Dynamics* as equivalent to a black-page Appendix.

## 20.5  Changes in Communities

In the preceding four sections, we have studied the structure and size of communities in terms of food supply and niches. Communities have an ordered structure and alteration of any part of this may have surprising effects. If we knew and could quantify all the factors affecting each species, then it should be possible to devise a computer programme which could be used to predict the effect of any change in any of these factors. Our present knowledge is very far from this ideal.

We know that communities change with time—this is called *ecological succession*. Take European lakes as an example: as a result of erosion of the valleys in which they lie, and deposition of the silt carried in inflowing streams, lakes gradually become shallower. The phytoplankton species change and, as the area of shallow water enlarges, the rooted plants increase in numbers and change in species. Eventually the shallow water turns into swampy marsh, shrubs begin to grow among the rushes, sedges and rank grasses, and finally, if left to itself, the lake basin turns into woodland. As the environment and plants change, so the animals change. The fishes of deep, rocky mountain lakes are trout and char, those of shallow, silted lakes are bream, tench and their relatives; marsh animals are different from those of lakes and woodland animals different from those of marshes.

**ecological succession**

This phenomenon of ecological succession is a natural process by which communities change gradually until they turn into the *climax*—the typical final community characteristic of that area. This may take thousands of years. The appropriate animals and plants are able to colonize the changing area because they are already present in adjacent areas. The climax community usually has many species and niches and is more complex than the preceding ones. Because of its ecological diversity, it is usually stable. Under some conditions, there is a cyclical change from the climax back to a slightly earlier stage in the succession and then on to the climax again.

**climax vegetation**

### 20.5.1  Rabbits

Since the species within the community interact, and there is a delicate balance between them, the introduction of a foreign species, or some change in a density-dependent factor affecting a native species, may have far-reaching effects on the rest of the fauna and flora. The rabbit is a native of Europe which has been carried to many parts of the world because it is good to eat and provides a useful fur. It breeds readily in captivity and is fairly hardy, being able to survive in central Africa as well as in the rigorous climate of Tierra del Fuego. It was probably introduced into Great Britain during Norman times. It is fairly closely related to the hares, of which there are two native British species.

Historical records show that rabbits were important in English feasts in the late thirteenth century and fetched high prices; by the sixteenth century, they were clearly abundant and rabbit warrens were maintained on various estates as sources of food and fur. In 1948, the annual turnover in wild rabbits was probably between 60 and 100 million.

The first rabbits probably reached Australia in the late eighteenth century, but experts believe that this introduction was not successful and that all the subsequent rabbits in the continent were actually the descendants of only 24 wild rabbits introduced into the state of Victoria in 1859. Rabbits were phenomenally successful and spread through some two-thirds of the

**rabbits in Australia**

continent. Foxes were introduced (for sport) in 1868 and have also spread through much of Australia—now, with a bounty on their heads, thousands are killed each year—but they did not control the rabbits. Trapping rabbits became a considerable industry with an export trade worth several millions of pounds annually of meat to Britain and fur to the USA.

What effects did this fast-breeding herbivore have on the native organisms of the countries into which it was introduced? In Australia, the unique mammal fauna (which will be referred to again in Unit 21) has suffered greatly, both from competition with and from predation by introduced mammals such as the rabbit and fox. Rabbits also competed with the introduced sheep and cattle. With their burrowing habits, they contributed to soil erosion.

The influence that rabbits had on the environment was shown clearly when their numbers were drastically reduced following the impact of the virus disease, myxomatosis, first in Australia and later in Britain. This disease was first reported in 1898 in Uruguay, where it caused an unusually high mortality of nearly 100 per cent among the European laboratory rabbits. Much later, in 1942, it was proved that the native species of wild rabbits in South America are immune to the virus, probably as a result of long association between this species and the disease. When the disease was transmitted to a European rabbit, this had no immunity and died. Colony and field trials with myxomatosis in Australia between 1936 and 1943 were only partly successful and the disease did not spread, possibly because foxes killed the infected rabbits. Six out of seven further liberations of diseased rabbits in 1950 were also limited in their effects, but the seventh led to rapid dispersal of the disease, with mosquitoes acting as carriers. Some million square miles were infected in the first summer; the disease died down in the winter but reappeared in epidemic form in the following summers; probably 1952–3 was the peak year. In the first few years, the mortality among infected rabbits was 99.5 per cent, but this fell off to less than 90 per cent by 1955—probably due to change in the virus and the appearance of antibodies in young rabbits. There were dramatic effects on sheep production—the increase in wool production alone was worth some 34 million pounds; other increases in rural production brought the total profit to 50 million pounds.

**myxomatosis in Australia**

The disease was introduced into France deliberately in 1952 and spread rapidly reaching almost every part of the country in 1953, and also the Netherlands, Germany and Spain. In Britain, wild rabbits with myxomatosis were first seen in the summer of 1953 in Kent and Sussex. The method of introduction is not known, but it was not a deliberate act by anyone in authority and at first there were attempts to contain the centres of infection with rabbit-proof netting. This became impracticable as other outbreaks started and, by the end of 1954, the disease had spread almost throughout the country. In Britain the carrier is the rabbit flea; rabbits which do not make burrows and nests are much less likely to contract myxomatosis than the others.

**myxomatosis in Europe**

Refer back to Unit 19, section 19.6.1, for more information about the disease and changes in the virus and in rabbits.

**What were the effects in Britain of the disappearance of almost the whole rabbit population through myxomatosis?**

First, consider changes in the vegetation: the areas where the disappearance of rabbits has made the greatest impact are the chalk downlands characteristic of southern England. These were grazed by sheep in the Middle Ages, but in more recent times the principal grazing animals have been rabbits.

43

Film strip 20(b) includes two pairs of photographs: one of each pair taken before and one after myxomatosis had killed almost all the rabbits in that locality. Both are chalk areas. Lullington Heath, Sussex, was photographed in early spring and Old Winchester Hill, Hampshire, was photographed in August.

**Look at film strip 20(b): numbers 5 and 6. What has changed between 1954 and 1967?**

In 1954, the rabbit burrows on Lullington Heath were surrounded by broken turf; there was much bare chalk. There were bushes near the burrows—these are elder (*Sambucus nigra*), plants which are unpalatable to rabbits.

plants of chalk downland

In 1967, there was no bare chalk exposed; the area was covered by a dense sward of grass (red fescue *Festuca rubra*) and there were bramble bushes (*Rubus sp.*).

Thus there has been an obvious change in vegetation. Grass and brambles grew in 1967, where there were few or none in 1954.

**Look at film strip 20(b): numbers 7 and 8. What changed between 1954 and 1956?**

In 1954, there was much bare chalk 'scree'; there was much ragwort (*Senecio jacobaea*), the tall plant with yellow flowers like those of daisies.

Two years later, in 1956, most of the bare chalk had been colonized by plants, mostly grasses. The photograph shows no ragwort plants.

**Suggest an explanation for the presence of ragwort in 1954 and its absence in 1956.**

Ragwort, like elder, is unpalatable to rabbits, so it was able to grow in places where rabbit grazing kept down palatable species. Each ragwort must grow from seed; it normally flowers and dies in the second year of its life (it is a biennial). Ragwort seeds established themselves on exposed chalk, where the ground was scratched by rabbits. After myxomatosis, plants of other species grew over the bare chalk soil and ragwort seeds failed to establish themselves in the face of this competition.

ragwort

The plants of the chalk downland of forty years ago usually formed a community with leaves close to the ground; often they had attractive flowers (such as thyme). Absence of grazing allowed faster growing shrubs and herbs with leaves further from the ground to spring up and over-shadow the traditional chalk downland plants. People who are anxious to preserve the original attractive community now work to clear the scrub of brambles and hawthorn, sometimes using mechanical mowers. In other places, sheep are being used for grazing again (at some cost!).

Secondly, the predators on rabbits—these are listed (in the *Handbook of British Mammals*) as: fox, stoat, buzzard, hawks, raven, crows, great black-backed gull, wild cat, farm and feral* cats and dogs, weasel, some owls and badger (the last three taking young rabbits and the rest taking rabbits of all sizes). Of these we shall consider only the fox.

Unfortunately, there are few good figures for fox diets before myxomatosis, but these suggest that rabbits and hares together formed more than half

fox diets

* *Domestic cats and dogs which have become wild.*

the mammals of the diet (unfortunately it is not possible usually to distinguish between these two as remains). A study of fox stomachs collected between 1955 and 1957 showed rabbits and hares in only about 20 per cent of stomachs; the most common small mammals were field voles, with immature brown rats next in frequency. Birds were found in nearly two-thirds of the stomachs from lowland foxes and the proportion of poultry and game birds in these stomachs was twice as high as before myxomatosis; hill foxes ate fewer birds and the occurrence of poultry and game was about the same as for earlier records. Hill foxes ate more sheep, probably often as carrion; the occurrence of these was about the same as for earlier records.

Foxes are carnivores eating a wide range of animals, including worms and beetles as well as birds and mammals; they also eat garbage and carrion. The disappearance of rabbits removed a prey species which had been an important component of their diet, but observers found no noticeable reduction in the weights, condition or breeding records of foxes in the next two years. Certainly their diets changed, but they were able to switch to other animals that were common and available, especially to field voles. Complaints of increased damage to poultry and game birds were probably justified in southern England but there is no evidence to support complaints of increased damage to sheep in hill districts.

Thirdly, competitors—the mammals most likely to compete with rabbits are their close relatives, the hares. Their habits are different: hares do not burrow, and they lead more solitary lives; but both species are herbivorous and were widely distributed. Hares are native to this country, whereas rabbits, as explained earlier, were introduced about 1 000 years ago.

hares

Here is some information for an estate of about $8 \times 10^6$ m² in the East Midlands: the first rabbit with myxomatosis was observed there in September in 1954; 1 800 dead rabbits were picked up in the eight weeks following this.

| | Number of rabbits killed annually | Number of hares killed annually |
|---|---|---|
| Before myxomatosis | 1 000 to 4 000 | <10 |
| After myxomatosis | | |
|    Winter 1954–5 | 0 | 10 |
|    Winter 1955–6 | 0 | 101 |

In September 1954, 62 hares were counted on the estate. In June 1955, a similar count gave 105 hares, including 9 leverets. In March 1956, at the end of the shooting season, 76 hares were counted on the estate.

Similar records have been published for estates in France. Where the annual 'bag' before myxomatosis was about 10 000 rabbits and between 100 and 250 hares, the numbers killed in the three years following the appearance of myxomatosis were less than 100 rabbits, but from 230 to 500 hares. The hare populations continued to increase slowly even when the number of rabbits began to increase again.

Biologists realized before 1954 that the activity of rabbits was important in determining the composition of certain communities. Enclosure with rabbit-proof netting of small areas of land was followed by the appearance in the enclosures of some plants that did not grow outside. But the almost complete disappearance of rabbits following myxomatosis led to some changes in communities that surprised most biologists. This is just one illustration of the fact that communities are in a delicate state of balance; alteration of part of the structure of a community may be followed by far-reaching and unpredicted changes.

Now you can attempt SAQ 14.

## 20.6  Man and the Environment

In pre-agricultural times, man ate berries and roots and killed various animals for meat. An omnivore with a wide food range can readily adapt its diet to changes in numbers or abundance of food organisms and so does not eat out any food species. The deliberate use of fire by man probably started quite early: this certainly alters community structure, favouring fast growing plants that can take advantage of devastated areas, such as rosebay willowherb, which is called 'fireweed', or plants resistant to heat and smoke. Large areas in Africa probably owe their present character to regular burning—and so do British grouse moors.

The development of agriculture implies very considerable interference with the natural environment. The climax vegetation is destroyed and replaced by a mono-culture—a field of wheat or cabbages or turnips replaces a forest of oak, ash, hazel and many other plants. The crops grown have characteristic demands on the soil. If the same crop is grown for several seasons in succession, the yield may fall off because some essential substance has been used up or because there is an accumulation of exudates from the crop plants. Such problems can be avoided by 'rotation of crops' or by adding the necessary substance—that is, by using fertilizers. We shall refer to this again in Unit 34. A very modern predicament arises from the use of fertilizers in such quantity that water draining from the land becomes charged with them, with consequent ill-effects on streams and rivers and on users of water.

agriculture

From the biological point of view, agricultural practice is completely contrary to the natural development of communities. Agriculture typically involves mono-culture—areas, often of considerable extent (some forests), planted with a single species of plant in place of the climax community of many species. As a result there emerge two types of biological problem organisms: weeds and pests (though the latter term may be used to include the former).

Weeds are often opportunist plants—usually quick growing and early flowering annuals*—naturally adapted to take advantage of the 'gaps' in normal communities, where soil is exposed and there is plenty of light. Other weeds appear to have evolved in parallel with cultivated plants; they have no known 'wild' habitats. Some weeds, such as many grasses, reproduce vegetatively** and so can cover wide areas quickly. If cultivated ground is left to revert to the wild, the weeds are ultimately replaced by other plants as the stable climax community develops.

weeds

Pests are animals which interfere with man's agricultural practices. Most of them compete with man by consuming the crop (or part of it) or making it useless (as some insects affect forest trees grown for their wood). Just as weeds are wild plants taking advantages of niches provided by man, so pests are wild animals with their own niches in wild communities. Many herbivores are selective in their food requirements. A natural community with many species of plants in the same area will support many species of

pests

* *Plants that flower and die within a year of the seed germinating.*

** *Without producing seeds; e.g. potatoes are propagated by tubers and strawberry plants by runners.*

herbivore in that area. For each of these species, there may be a problem in finding food plants at certain stages of their life cycles—these plants may be some distance apart and the animals need some form of dispersal. The herbivores are preyed upon by carnivores, some of which are very specific in their choice of food, and are attacked by parasites, which are often specific in their choice of host. In a long established community, there will be a stability in the balance between all these organisms, although they will probably show fluctuations in numbers.

Agriculture provides enormous food resources for those herbivores which share the same tastes as man. With many plants of one species grown close together, there is no problem of individuals finding new food plants at the dispersal stage. Thus it is possible for very large numbers of herbivores to build up on the crop plant. You may wonder why the predators and parasites of the herbivorous pests do not also build up in numbers. In fact, this must start to happen very often but, if the predators or parasites keep pace in their numbers with the herbivore, then the latter's numbers are always kept at a relatively low level and so by definition it is not a pest. Wherever crops have grown apparently free from pests, natural regulation must have operated. But if the parasites and predators are not fully synchronized with their prey, or have some special requirement during part of their life-history (special conditions in which to hibernate or lay eggs or rest during the midday heat), they may not be able to follow their prey into the cultivated land. Slight changes in the pattern of cultivation may accommodate the predators or, on the other hand, make it impossible for them to act as regulators. The removal of hedges may well reduce the numbers of predators, such as birds, beetles and hedgehogs, that keep the numbers of pests and potential pests at low levels in smaller fields.

A different problem arises when some animal is introduced into a new area where there is a suitable food supply. The introduced pest may lack completely its specific parasites and it may not be favoured by the local predators. Under these conditions it is able to build up large numbers without any regulation whatever. With air transport all over the world, the chances of casual introduction of new pests are much greater than in the days of steamships or sailing ships. Winter moth in Canada (20.4) is an example of an introduced pest with no natural enemies in Canadian forests.

## 6.1  Pest control

Since pests are animals present in large numbers, they clearly lack natural regulators to keep the populations at low (therefore agriculturally tolerable) levels. One approach to the problem is to try to introduce appropriate regulators—this is called *biological control*. Another approach is to use chemical substances (pesticides) to kill the pests—this is *chemical control*.

**Two examples of biological control have already been mentioned in this text. Name them.**

The introduction of parasites of winter moth from Europe to Canada. The introduction of myxomatosis to control rabbits in Australia and France.

Both these examples illustrate the essential point about this method: the introduced agent (parasite, predator or disease) must be specific to the pest. If it also attacks other organisms, then the effects of introducing it are very difficult to predict and may turn out to be disastrous in other ways. This happened when the mongoose was introduced to control snakes in Haiti—having eaten out the snakes, the large populations of mongoose then turned their attention to many species of birds and mammals. This might have been predicted since mongoose are not specific

in their feeding habits and will eat birds and mammals as well as snakes.

The use of chemical control is often rather similar to the introduction of non-specific predators—but has the advantage that it is easier to stop. Most pesticides are not specific to the particular pest but kill a large range of other insects. In field conditions, these other insects will probably include predators and specific parasites of the pest—that is, the natural regulators. If these have been killed completely, then the only way of controlling the pest is to continue to use the pesticide year after year. Sometimes the indiscriminate killing of predatory insects means that other herbivores, which were being regulated naturally, become pests. To have survived, these must be at least slightly resistant to the pesticide; thus a new problem has appeared which may well be worse than the original infestation for which the pesticide was used. An example of this, quoted in Unit 19, is the appearance of flies resistant to DDT.

chemical control

The case of the cottony-cushion scale in California illustrates some features of biological and chemical control. Turn to p. 48 of *Population Dynamics* and read the first paragraph of 7.3; the lower photograph of Plate 1 shows the scale and the introduced specific predator, the ladybird *Rodolia*. After the War, citrus growers started to use chlorinated hydrocarbon insecticides such as DDT. One result was the re-emergence of cottony-cushion scale as a pest in Californian orchards, since the pest seems to be much more resistant to these insecticides than is the ladybird. When the citrus growers altered their insecticide programmes, *Rodolia* was able to survive and to keep the scale under control. Note that the ladybird used to be called *Vedalia* and you may find it described under that name in some books.

biological control

While biological control is clearly an elegant way of controlling pests— and the most economical way from the point of view of running costs—it must be realized that the search for appropriate regulators may be long and may not always be successful. Insecticides are necessary adjuncts to agriculture, but it is essential that they be used discriminatingly and with a clear understanding of all the possible biological consequences. This has not been the case in the past. When the choice of insecticide lay between substances such as derris,* which break down quickly and are not toxic to mammals or birds, or such obviously dangerously poisonous substances as arsenates or lead salts, the use of insecticides did very little real damage to the environment. The situation was changed by the development of DDT and related compounds. These are very effective and persist for long periods in the soil, so they appeared at first to be ideal as agricultural pesticides. It was some years before it was realized that they act as cumulative poisons in birds and mammals and are passed along the food chain and so accumulate in carnivores. The battle to restrict the use of these stable, persistent pesticides is still raging; you should find it of interest to read *Silent Spring*, the book by Rachel Carson that first made the general public aware of the very dangerous situation arising from the indiscriminate use of these substances. *Pesticides and Pollution* is a more recent account of the problem.

DDT

Another type of 'chemical' control of pests is through the use of *systemic insecticides*—substances taken up by the plant and poisonous to the herbivore feeding on it, but not affecting any other animals. Analogous with this is the use of bacteria or viruses as insecticides—these, again, are highly specific and, although essentially 'biological', can be treated as chemical substances, stored and used as dusts, as and when convenient or necessary.

systemic insecticides

The use of sterile males is a different sort of biological control which has been successful with the screw-worm fly in the southern U.S.A. The theory is that the females which mate with the sterile males will lay eggs which

* *An insecticide extracted from plants; the active principle is rotenone.*

fail to develop. If sterile males form a large proportion of all males present, then a large proportion of the eggs must fail to develop. The practical problem is to raise sufficient numbers of males (which are then sterilized by exposure to radiation) to make any impact on the population; the effort required is worth while only when a major pest is involved and there is no easier safe method of control.

The future is likely to see a balance between chemical and biological control, with the use of each method where appropriate, and much greater care taken not to cause indiscriminate mortality of pests, predators, parasites and insects irrelevant to the problem. With a better knowledge of the life-histories of the organisms involved and improved understanding of the theories of population dynamics, it should be possible to use computers to work out the likely effects of different programmes and so to choose the best one.

A curious twist to the use of biological control is the recent development of a new method to control red spider mite on tomato plants in greenhouses. Red spider mite became a pest after the introduction of DDT, **red spider mite** since it is very resistant to this substance and to all other pesticides. It can be regulated at a low level by a predator. To ensure that the predator will survive to exercise its regulation, the new technique is to introduce some red spider mites at the same time. Once the predator and prey populations are established, they will both keep going at a low level; this is low enough for the tomato plants to fruit freely and successfully. Previous efforts to use the predator sometimes failed because the predators died out before they had found enough prey to become established!

## 9.6.2    Human populations

Turn to *Population Dynamics*, p. 53: read section 7.8, then turn back to this text.

The rapid increase in human population since the seventeenth century has inevitably affected other species of organisms. The old agricultural populations have been replaced in certain parts of the world by communities founded on industry and technology (refer back to Unit 1 and to *The Roots of Present-day Science* for more details). Some of the effects of man on his environment can be classed under the general heading of 'pollution', **pollution** and we shall refer to this again in Units 33 and 34. With increasing awareness of the problems and with improved technology, most pollution could now be reduced or prevented—but at a cost. The problems become economic and political. How can a politician balance the value of an amenity such as a clean river, where anglers can catch fish and people can swim or boat or simply enjoy a walk along the riverside path, against the thousands of pounds it may cost industry and the community to improve industrial and sewage effluents? Or the value of woods in city suburbs, where children and adults can 'enjoy nature', against the value of the site in terms of land for building houses or factories—or even motorways? These are the sort of problems where the conservationists must argue their case against industry and other cost-conscious sections of the community; there is usually no dispute about the scientific aspects of the ecological situations involved.

Other effects of man on the environment arise simply as a result of the increase in human numbers. These are problems of food and of living space, problems which can affect all species of animals. The second of **crowding** these problems has rather been overlooked so far except in terms of the first (that is, that the increasing population will mean less land for food production); but it has been suggested that crowding may itself affect the

human population as it affects rats and voles when these animals are unnaturally crowded both in experiments and as the result of peak populations in the wild. In these rodents, stress symptoms appear; the adrenal glands are enlarged and the birth rate is reduced. Crashes in wild animal populations can also result from epidemic disease or from eating out of some important food resource. Human technology may be able to overcome these two last possibilities but, if there are hormonal effects of crowding, these may act as population-limiting devices.

### 20.6.3 Human food supplies

The food supply problem has been much discussed, though there remain two contrasting points of view. Many agriculturalists optimistically believe that, with the use of fertilizers, herbicides and pesticides, the world can produce enough food even for the enormous population predicted for the next century. On the other hand, many ecologists and others, knowing that at least half the population of the world is at present underfed, do not believe that it will be possible to produce enough to feed adequately the probable population of thirty years ahead. The estimates of these two groups of scientists differ widely; here we can only draw attention to the sort of problems which are relevant.

**estimates of food supplies**

If you turn to *Ecological Energetics*, p. 45, Figure 5.1, you can see how the proportion of the world population which is underfed has increased as the total population increased.

There are three main methods of improving this state of affairs: first, to stabilize or reduce the human population; secondly, to increase the efficiency of production and utilization of the present food sources; and, thirdly, to develop new food sources.

At present there is a high loss of food during transit from its source of production to its final utilization. Some of this loss of stored products occurs through pests—mice, beetles, flour moth and moulds—and could be avoided. Further research on the pests of stored foods should make the utilization of what is at present produced more efficient. Some food is lost for economic reasons such as the high cost of transport from areas of over-production to places where food is scarce; these political problems require political solutions. It seems on the face of it nonsense that producers should be burning food, or allowing it to rot, in one part of the world when people are actually starving or on the verge of starvation elsewhere.

The problems which we can reasonably discuss here are those of the efficiency with which different types of food can be produced in different environments and the likelihood of producing enough to fill the bellies of the future world population.

**food production**

Since energy is lost at each link of the food chain, the most efficient way for an omnivore like man to tap solar energy is to eat plants. Turn to p. 46 of *Ecological Energetics* and look at Figure 5.2. This shows the primary production in various types of ecosystem through the world. The highest figures for cultivated plants come from tropical agricultural crops such as sugar cane, which grow all the year round; for most agricultural crops the average annual production works out as between 3 and 10 $g/m^2/day$. Turn to p. 37 of *Ecological Energetics* and look at Table 5—figures for annual production of various plants are given as metric tonnes/hectare. How is this related to the values of production in Figure 5.2?

1 tonne $= 10^3$ kg $= 10^6$ g; 1 hectare $= 10^4$ m$^2$; 365 days/year so divide the values in Table 5 by 3.65 to make them comparable with the values in Figure 5.2. Notice that both sets of figures are for dry weight. What sort of conversion factor would you use to turn these into wet weights?

As you read in Unit 14, water comprises at least 60 per cent of most living organisms. For realistic values, you should probably multiply the figures by at least 3 to obtain the primary production in wet weights.

Most of the plants in Table 5 are not eaten by man. Usually only parts of crop plants are eaten by herbivores—the fruit, seeds, leaves or roots, but not all of them—so much of the primary production is not utilized. Plant proteins differ from animal proteins in proportions of amino acids; herbivorous animals have to consume large amounts of food simply to obtain certain essential amino acids. Consuming an animal diet is economical in terms of bulk of food to be consumed, although it is wasteful of solar energy. About 70 per cent of the present supply of protein for human food comes from vegetable sources and about 30 per cent from animal sources.

The majority of domesticated food animals are herbivores. But, as converters of solar energy into food available for man, there are marked differences between different herbivores. Figure 5.4 of *Ecological Energetics* illustrates the contrast between cattle and rabbits; the rabbits produce the same weight of meat from a given amount of hay as do beef cattle, but they do so in 30 days instead of 120, so their production is more efficient per unit time. Intensive livestock rearing is the most efficient way of producing those animal proteins which are traditional foods. The chickens and calves are harvested when young and still growing fast; they are fed on processed diets which incorporate parts of plants (and animals) normally discarded as inedible. Sometimes unfortunately, substances are added to their diet that may be dangerous to people, such as, for instance, antibiotics and hormones.

**domesticated food animals**

Ideally, protein should form about 10 per cent of total food intake, measured in terms of energy. The really big difference between the diets of well-fed and under-fed peoples of the world is, in fact, in the amount of protein eaten. An adequate diet should contain about 44 g/day: in North America, the present average consumption is about 64 g/day whereas in underfed areas it is about 9 g/day.

Apart from the traditional domesticated animals, there are two main sources of animal protein—wild mammals and fishes.

Wild mammals are principally of interest in the tropics and in the polar regions (seals, whales and polar bears). There is a strong case for harvesting wild game in parts of Africa, instead of introducing domesticated cattle and goats. The wild game are adapted to the natural ecosystem and the many species of antelope together with zebra, rhino, elephant and hippo presumably rely on slightly different plants and therefore can all flourish together. Introduced cattle eat only a limited range of plants and are not adapted to the local climate nor capable of withstanding the local diseases.

**wild mammals**

Sea fisheries depend on harvesting natural populations of fish. Some of the problems involved are discussed in this Unit's TV programme. Optimum exploitation, maintaining a good yield without damage to the breeding stock, depends on careful manipulation of the fishing effort so that the numbers and sizes of fish caught are regulated. The scientific principles basic to this rational exploitation are well understood, but their application depends on international co-operation—so far this has been achieved only in very few cases. Managing wild populations of whales, seals and polar bears depends on exactly similar principles to those basic to the fishing industry. The present disastrous state of whaling in the Antarctic is a sad example of the effects of over-fishing, carried out in the face of repeated warnings by marine biologists, and in spite of the fate of whales in northern

**sea fisheries**

waters. Many of these marine sources of first class protein are carnivores—a new threat to them is the pollution of the sea by substances which pass along food chains and accumulate in top carnivores. Insecticides such as DDT, fungicides containing mercury, PCBs (poly-chlorinated-biphenyls) used in industry, and other substances can reach the sea through industrial and sewage effluents or run-off from the land; their accumulation in fishes could affect disastrously the potential food supply for the rising world population.

In 1962, there were 1.3 acres of arable land per person, and it is estimated by FAO* that the figure will be 0.7 acres per person by 2 000 A.D. This implies that merely to maintain the present levels of nutrition by agriculture alone would involve doubling the productivity of the arable land—a very unlikely prospect. It is possible that the production from the sea could be doubled and production of terrestrial animal protein might be increased by rather more than 50 per cent; there are possibilities of increasing freshwater fish production by more than 10 times. Between them, these actions would just about double the amount of animal protein available in 2 000 A.D.

There are other possible sources of food: proteins can be extracted from leaves to provide a nutritious diet lacking only the essential amino acid lysine; a bacterium converts petroleum into protein which can then be harvested; biochemists could produce synthetic foods. All these sources can be exploited only through manufacturing processes which use energy; the amount of energy necessary to produce the edible protein may be so great that the source is totally uneconomic. Indeed, since the manufacture of synthetic fertilizers requires expenditure of energy, the use of increased quantities of fertilizer means faster consumption of the world's reserves of fossil fuels, as also does the increased use in agriculture of insecticides, herbicides and other synthetic substances. The machines that apply these substances to the land and the mechanical harvesters that collect and process the crops also use considerable quantities of fossil fuel. Feeding the world population uses up a great deal of energy—the reserves of fossil fuels are being depleted rapidly.

*new sources of food*

*energy requirements of food production*

Nuclear energy, or some process for using the solar energy at present wasted, will be essential if a large human population is to survive in the future. A new source of protein obtained by a process based on one of these types of energy could make it feasible to feed three or more times the present population of the world. The limit to population increase might then be set by density-dependent physiological or psychological reactions. We can observe these operating for other species of animals, but at present we know very little about such reactions in human populations.

*Now you can attempt* SAQs 15 *to* 19. *These relate to the whole Unit.*

If you wish, you can read *Ecological Energetics*, chapter 5, and *Population Dynamics*, the rest of chapter 7, as black-page Appendices to section 20.6.

* *The Food and Agriculture Organization of the United Nations.*

## 20.7  Summary

All individual organisms belong to assemblages of individuals called communities; these usually include individuals of many different species. Individuals and species are not randomly distributed. The composition of any given community can be understood in terms of food webs and of the 'flow' of energy and of essential elements through it; these can be represented as diagrams and also as several different types of ecological pyramid.

Each species in a community occupies a 'niche', defined by its food relations and life-history. The geographical distribution of species depends on where they evolved, how individuals can be dispersed, how tolerant individuals are of various environmental conditions summed up as climatic and edaphic factors, and how successfully individuals can find food resources and can survive in the 'struggle for existence' (in the face of predation, parasitism and competition).

Studying population numbers involves careful sampling techniques. It is important to measure the fecundity (or numbers of young produced) and the mortality factors operating at different ages or stages. Survivorship curves and curves of population change with time show considerable differences between species. Analysis of mortalities using $k$-values reveals which are the 'key factors', mainly responsible for changes in numbers, and which are the 'regulating factors' that tend to stabilize the population by altering the numbers in the direction of the average values. This type of analysis is described for an insect (winter moth) in the text and for a bird (tawny owl) in Appendix 1.

Natural communities change with time towards a fairly stable 'climax'. Interference with communities may cause considerable changes, not fully predictable at our present stage of knowledge. The impact of the virus disease myxomatosis on rabbit populations in Australia and in the British Isles, and the consequent changes are examples of effects on communities resulting from human actions.

Human civilization has considerable effects on other organisms; the emergence of weeds and pests and some methods of pest control are studied as special examples. Finally, some of the problems of the expansion of the world human population are discussed with special reference to food and energy requirements related to energy flow in different types of community.

# Appendix 1

## Tawny Owls

The information in this Appendix is taken from a study of tawny owls on the Wytham Estate, near Oxford. For the full account, see: H. N. Southern (1970) 'The natural control of a population of tawny owls (*Strix aluco*)' *Journal of Zoology*, Vol. 162, pp. 197–285.

### 1.0   The life history of the tawny owl

The tawny is the most common owl species in Great Britain, where it is widely distributed in woodlands, open ground and cities, wherever there are some large trees. It roosts in trees by day and becomes active at dusk; its method of hunting is to perch on a low bough of a tree and then drop on to its prey, which it probably detects by the sounds made. This species of owl nests in holes in trees—thus large trees are absolutely essential for its way of life.

Tawny owls are highly territorial (see 20.4). They probably spend their whole adult lives in one area. They have characteristic calls so that it is possible to locate the birds and to recognize what they are doing at certain times and seasons.

A typical year in the life of a pair of tawny owls is as follows.

*August and September*   The birds moult (i.e. they lose their feathers and grow a new set); at this time they are silent and inconspicuous.   **moult**

*October and November*   Each pair of birds asserts its territory. Each owl spends 15 to 20 minutes at dusk hooting (hōoo-hu-hŭ,-hŭ,-hŭ,-hōoo) from a prominent tree. Later the owls hunt for food. If an owl encounters an owl other than its mate, it challenges noisily by 'caterwauling'. The limits of the territories are established by noisy encounters between the residents. Each resident owl is reluctant to leave its own territory.   **establishment of territory**

*December to February*   The owls begin to indulge in courtship behaviour. They choose a nest site, a hole in a tree; they roost together, often near the nest site. The male brings food to the female who calls softly (oo-wip) to attract him.

*March to May*   This is the breeding season. The female lays up to four eggs and begins to incubate (to sit on them) after she has laid two. She stays on the nest, leaving it only when the male brings her food. When the eggs hatch, the female stays with them until they are halfway to being 'fledged' (able to fly); she tears up food brought by the male and distributes it to the owlets. When they are half grown, she starts to help the male to collect food for them.   **breeding**

*June to August*   The owlets become fully fledged and leave the nest. They explore their parents' territory, but still rely entirely on their parents for food. Owlets call (ti-swerp or ti-swoop) at dusk and through the night; this shows their parents where to find them with food.   **rearing young**

*August and September*   The owlets disperse and must start to fend for themselves. Some pair, find territories and may breed in the following year. The adults start to moult and the annual cycle is then repeated.

*Question 1*   Using the information given above, suggest possible methods and times for carrying out censuses of the tawny owl population in a given area.

*Read Answer 1 (p. 62).*

## 1.1   Territories of tawny owls on the Wytham Estate, Nr. Oxford

Southern started his study in 1947, after a winter of prolonged frost and snow when many adult tawny owls died probably from starvation. In that autumn, there were only 17 pairs of owls occupying territories on the estate. The numbers increased gradually; in the five years from 1955 to 1959 (the end of the study) there were 30, 32, 32, 31 and 32 territories, suggesting that 32 is the upper limit to the viable number on this estate. The study therefore covers a period when the density of pairs of owls was increasing towards and then settling down at this maximum value. We shall refer to this later.

The estate, of total area $525 \times 10^4$ m², includes dense woodland, open woodland and parkland. In the later years, the mean area of each territory was $16 \times 10^4$ m², but territories in open parkland were always larger than those in woodland. The boundaries of some territories remained remarkably constant from year to year. The same pair of owls was known to occupy certain territories for seven years or longer; on average, each pair of owls occupied a territory for five years. In the first year or two, the pair often failed to rear young, but usually they bred successfully in the later years of their occupation of the territory.

**owl territories**

*Question 2*   Suggest an explanation for the observations in the last sentence.

*Read Answer 2 (p. 62).*

*Question 3*   In the light of these observations, suggest a possible survival value of the owls' territorial habit.

*Read Answer 3 (p. 62).*

## 1.2   The food of tawny owls at Wytham

Owls are carnivores, but there is seasonal variation in their diet.

Between December and February, about 80 per cent of their diet consists of small mammals, principally wood mice and bank voles. From May to August, moles, young rabbits and invertebrate animals such as earthworms and beetles become the principal items in the diet. Over the whole year, mice and voles comprise about 50 per cent of the vertebrate animals of the diet; these animals can be caught alive in traps, marked and released so that they can be counted. Here are some facts about them:

(a) The wood mouse *Apodemus* is strictly nocturnal in habit. By day it lives in burrows; it comes out at night and forages over the open woodland floor eating mainly seeds but, especially in spring, also some insects. It

**wood mouse**

breeds between April and October. Its numbers are least in early spring and greatest usually in the autumn.

(b) The bank vole *Clethrionomys* may be active at all times of day and night, but each individual probably has a rhythm of activity and rest. It feeds on live and dead herbaceous material, and usually moves about under shrubs and avoids open spaces. It breeds at the same time as the wood mouse, but may continue longer into the winter. The numbers are usually low in May and June and reach a peak in autumn or winter.

bank vole

Of the other animals in the diet: moles eat insects and earthworms; earthworms eat dead leaves; the beetles eaten by owls include cockchafers, which are herbivorous, and ground beetles, which are carnivorous, eating other insects.

*Question 4*  From the information given above, construct a food web for tawny owls at Wytham.

*See Answer 4 (p. 63).*

Figure 11 gives the results of censuses of wood mice and bank voles carried out at intervals of six months. The scale is arithmetic.

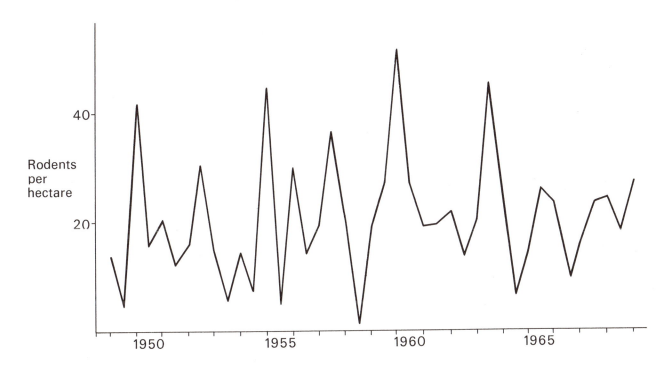

*Figure 11  The numbers of small rodents (wood mice and bank voles) estimated at six-monthly intervals in Wytham Wood.*

*Question 5*  How are the highest and lowest numbers of these small rodents related to each other?

*Question 6*  Do the fluctuations appear to follow a regular cycle?

*Question 7*  What, approximately, is the average number of small rodents per hectare ($=10^4$ m²)?

*Read Answers 5, 6 and 7 (p. 63).*

The detailed figures show that the greater part of the fluctuation in numbers is of numbers of bank voles. Wood mice seldom exceeded 10 ha⁻¹; the totals for the two species fluctuated about an average of 20 ha⁻¹; on three occasions, the numbers of bank voles approached 40 ha⁻¹.

*Question 8*  Suggest a hypothesis connecting these observations on bank voles and wood mice with the numbers of tawny owls.

*Read Answer 8 (p. 63).*

## 1.3   Life tables for tawny owls at Wytham

If tawny owls are provided with suitable boxes, they will nest in them, so that it is possible to observe whether a given pair lays eggs or not, whether the eggs all hatch and whether all the chicks are fledged. Chicks can be marked with rings and some information about survival of fledged chicks can be obtained from analysis of when and where these rings are recovered.

Here is the life table for tawny owls at Wytham:

*Average number of eggs laid*: 62

| Age in years | Average number surviving at the end of the year |
|---|---|
| 0–1 | 19 |
| 1–2 | 9 |
| 2–3 | 7 |
| 3–4 | 6 |
| 4–5 | 5 |

survivorship

*Question 9*  Construct a survivorship curve for tawny owls at Wytham.

Here is part of the detailed information collected.

| | *Data for 1952* |
|---|---|
| Number of pairs of owls present on the estate | 24 |
| Number of pairs attempting to breed | 17 |
| Average number of eggs laid per nest | 2.5 |
| Eggs lost before hatching (as percentage of total) | 62 |
| Chicks lost between hatching and fledging (percentage of total) | 6 |
| Average annual mortality of adult owls (percentage of total) | 18.5 |
| Number of pairs of owls present in the following year (1953) | 24 |

*Question 10*  Use the figures given above to construct a survivorship curve for the eggs and young of 1952.

*Look at Answers 9 and 10 (p. 64).*

## 1.4 Key factor analysis for tawny owls at Wytham

The average maximum clutch size was 3 eggs per pair, so the maximum number of young that are likely to be produced each year can be derived by multiplying the number of pairs in the area by 3. This maximum was never achieved during the period of observation.

The following mortality factors were quantified: (Refer back to 20.4 for definition of $k$-values).

$k_1$  the number of eggs 'lost' through failure of the adults to breed; (the $k$-value is the log of the number of pairs failing to breed multiplied by 3).

$k_2$  the number of eggs 'lost' through failure of those pairs that nested to lay the maximum clutch of eggs—the average clutch size was 2.5 eggs per breeding pair.

$k_3$  the number of eggs lost before hatching. Most of this loss was the result of desertion of the nest by the female.

$k_4$  the number of chicks lost before fledging. Most of these died either before their eyes opened or while the feathers were growing.

$k_5$  the number of young dying between leaving the nest and acquiring territories of their own in the following spring.

Figure 12 shows the $k$-values derived from the data for 1949 to 1959 inclusive.

Examine this graph, then answer the following questions.

*Question 11*  Which is the 'key' factor?

*Question 12*  Do any other $k$-values vary in the same way as the key factor?

*Question 13*  Do any $k$-values appear to vary in the opposite way from the key factors? (that is, are any of the $k$-factors compensating, density dependent factors?)

*Read Answers 11, 12 and 13 (pp. 64–5).*

The key factor analysis has shown which mortalities are responsible for the fluctuations of the population numbers and which mortality acts in a way that should stabilize the numbers. Can we relate these mortalities to any environmental factors that vary at Wytham?

*Question 14*  Consider what you have already read about tawny owls and then suggest environmental factors that are worth detailed evaluation.

*Read Answer 14 (p. 65).*

## 1.5 Mortality and food supply (in the form of bank voles and wood mice)

Probably adult owls died of starvation during the long snowy winter of 1946–7, just before the beginning of these observations, but Southern found no evidence of *adult* mortality varying with food supply between

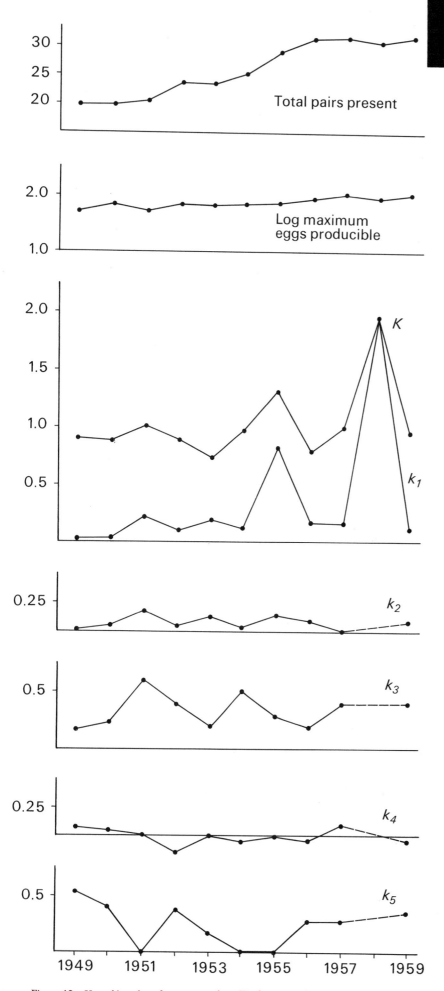

*Figure* 12 K *and* k-*values for tawny owls in Wytham Wood.*

1947 and 1959. The first hypothesis of Answer 8, thus, was not supported by the results of this investigation.

Figure 13 shows the relationship between $k_1 + k_2$ and the density of small rodents in spring (expressed on a logarithmic scale).

Examine Figure 13.

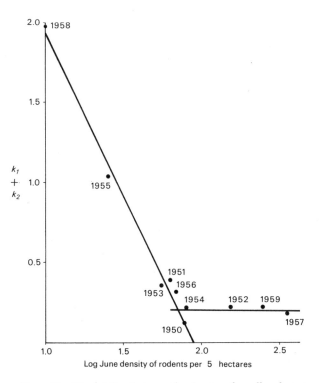

*Figure* 13  *The relation between the density of small rodents in June and 'losses' through failure to breed and to achieve maximum clutch size* ($k_1 + k_2$).

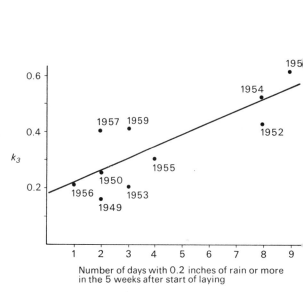

*Figure* 14  *The relation between rainy days and the failure eggs to hatch* ($k_3$).

*Question 15*  In a few sentences, state your conclusions from it.

*Read Answer 15 (p. 65).*

## 1.6  Mortality and weather

Figure 14 shows the relationship between $k_3$ and the number of days with 0.2 inches or more of rain in the 5 weeks after the owls started to lay eggs.

*Question 16*  Examine Figure 14 and state your conclusions from it in a few short sentences.

*Read Answer 16 (p. 66).*

## 1.7 Mortality and the density of the population

Figure 15 shows $k_5$ plotted against the number of young fledged in that year (on a log scale). The negative values for $k_5$ indicate that owls migrated into Wytham from other areas; this happened in years when an insufficient number of owlets survived on the estate to replace the adults that died.

*Question 17* Does Figure 15 indicate that $k_5$ acts as a density-dependent factor?

*Question 18* If you answered 'yes' to Question 17—Does $k_5$ act as a compensatory 'regulating' factor?

*Read Answers 17 and 18 (p. 66).*

Figure 16 shows the relation between the average number of young fledged per territory and the number of adult owls (expressed as the number of territories on the estate).

*Question 19* Examine Figure 16 and state your conclusions from it in a few short sentences.

*Read Answer 19 (p. 66).*

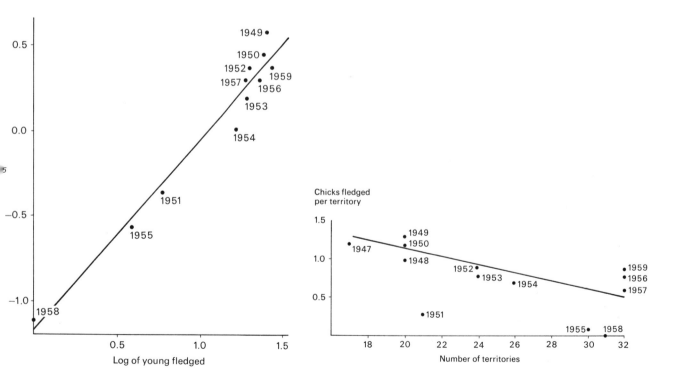

*Figure 15 The relation between the number of young fledged and the overwinter disappearance of yearling owls ($k_5$).*

*Figure 16 The relation between the number of territories held and the number of young fledged.*

Finally: write a short summary listing the factors that control the population of tawny owls at Wytham. Compare your summary with that in section 1.9, p. 66.


## 1.8 Answers to the questions

*Answer 1* Each pair of adult owls stays in its own territory, so it is possible to map the territories over a period of weeks and deduce the number of adult birds. Although the owls are nocturnal and fly silently, their calls allow observers to locate and count them.

*October to November* Hooting at dusk (Southern organized a large band of helpers on one evening each year to locate as many owls as possible during the 20 minutes after dusk when the owls called). The 'caterwauling' at the edges of territories allows them to be mapped. Owls roused from their roosts during the day refused to cross territorial boundaries and would 'bounce back', to use Southern's graphic description, marking the edges precisely.

*December to March* The females' calls during courtship, mating and nesting allow the pairs to be located and counted.

*June to early August* The calls of the young allow them to be located and counted.

Owls nest in holes in trees, but will use a suitable nesting box, which can be fitted with a mirror. They cannot 'see' red light and can be observed at night using a red torch. It is possible to locate many of the nests and to count the numbers of eggs laid and the numbers of young hatched and fledged. Food brought to the nest can be observed and identified. (At other times of year, the food can be deduced by collecting pellets from under the trees in which the owls roost when they regurgitate indigestible parts of their food—small mammals can be identified from bony parts and some insects from their hard wings or heads.)

Juvenile birds leave their parents' territory in August so 'ringing' individuals can be a census aid—a numbered metal band is fixed loosely round one leg; the numbers are recorded in a central register run by the British Trust for Ornithology, British Museum (Natural History). Anyone finding a bird with a numbered ring informs this registry and movements of birds can thus be mapped and some information gathered about the way in, and age at, which they die.

*Answer 2* The owls feed entirely within their territories; their expertness at locating and collecting food increases as they know the territory better. Success in rearing young means that the male bird has been able to bring sufficient food to the nest for the female to feed while she incubates the eggs and looks after the small owlets and also for the owlets to grow and become fledged. An inexperienced male may not be able to find sufficient food for himself and his mate, but the pair are more likely to be successful at rearing young in the later years of their occupation of the territory as the male becomes more expert.

*Answer 3* The habit spaces out the owls, giving each pair an area in which they can seek food without interference or competition from other owls. Since the population survives, the areas of the territories must be adequate as sources of food for a pair of birds and their offspring in

**Appendix 1**

average years. The information from Wytham suggests that smaller territories suffice in woodland compared with parkland—on this estate an average of 16 ha seems to be the minimum area that is adequate for survival.

*Answer 4*

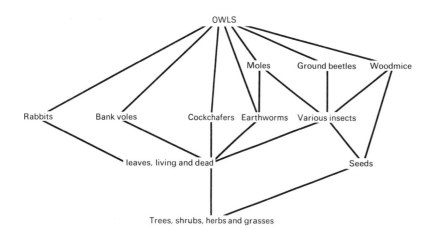

*Figure* 17   *Food web for tawny owls in Wytham Wood.*

*Answer 5*   The lowest figure recorded is that for winter 1958—approximately 1 small rodent ha$^{-1}$. The highest figure is that for spring 1959—approximately 50 ha$^{-1}$. So the highest value is about 50 times the lowest.

*Answer 6*   No, there is no regularity.

*Answer 7*   Approximately 20 ha$^{-1}$.

*Answer 8*   Bank voles and woodmice together form a substantial part of the diet of tawny owls, so fluctuations in numbers of small rodents may lead to fluctuations in numbers of tawny owls. About 80 per cent of the diet in winter consists of the two rodent species, whereas the diet in summer includes more other mammals and invertebrates than bank voles and wood mice. Here are two possible hypotheses.

1   When the rodent population is low in autumn and winter, the survival of tawny owls through the winter will be low—there will be fewer adults in the following spring.

2   When the rodent population is low in the winter and early spring, the breeding success of the tawny owls will be below average—since the males will find it difficult to bring enough food for the females to lay and incubate eggs in March and April. Females without sufficient winter food also may not be able to produce as many eggs as those with plenty of winter food.

*Answer 9*  See Figure 18. Starting with 62 eggs, the numbers of survivors are plotted as quoted in the life table.

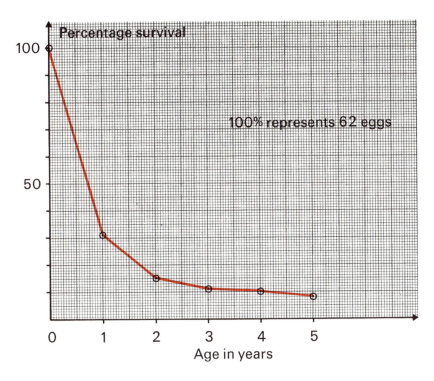

*Figure 18  Average survivorship curve for tawny owls.*

*Answer 10*  The information needs manipulation before the curve can be plotted. The figures needed are as follows:

(a) Total number of eggs laid (in March).

(b) Total number of eggs that hatch into chicks (in April).

(c) Total number of chicks that become fledged (in June).

(d) Total number of young owls that survive to form pairs in the following March.

To obtain (a) multiply the number of pairs attempting to breed by the average numbers of eggs laid per nest: $17 \times 2.5 = 42.5$.

To obtain (b) multiply the number of eggs laid by the percentage survival of eggs ($100 - 62$ per cent): $42.5 \times 38 \times 10^{-2} = 16$.

To obtain (c) multiply the number of chicks hatched by the percentage survival of chicks ($100 - 6$ per cent): $16 \times 94 \times 10^{-2} = 15$.

To obtain (d) note that the number of pairs of owls in 1953 was the same as that in 1952 (24 pairs). But 18.5 per cent of the adult owls died during the winter. Thus the number of owlets that survived to form pairs in 1953 must equal the number of adult owls that died during the winter: $24 \times 18.5 \times 10^{-2} \times 2 = 9$.

These figures are plotted to give the curve in Figure 19.

*Answer 11*  $k_1$ is the key factor. In 1958 it accounted for the total loss of recruits to the population.

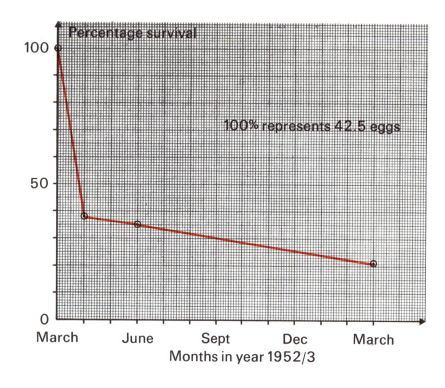

*Figure 19  Survivorship curve for the eggs and the young of 1952.*

*Answer 12*  $k_2$ varies in the same way as $k_1$, but is not so large. $k_3$ varies in a similar way to $k_1$ in some years (e.g. 1949–52) but in a quite different way in other years (e.g. 1953–55).

*Answer 13*  $k_5$ appears to vary in the opposite way from $k_1$ so that it tends to restore the level of the population towards the average value from extreme high or low values.

*Answer 14* (a)  The English climate is very variable so that it might be worth evaluating the relation between $k$-values and temperature and rainfall.

(b) Variations in numbers of small rodents might lead to fluctuations in survival or in breeding success of owls (see answer 8).

(c) The number of pairs of owls on the estate rose steadily from 1947 to 1955 (see section 1.2, p. 55). $k_1$ is lower in the early years than it is in 1955 or 1958; it might be worth investigating the relation between $k$ values and density of adult owls.

*Answer 15*  $k_1+k_2$ represents the 'losses' resulting from failure of some pairs of owls to breed and failure of those that bred to lay the maximum number of eggs. These losses are very high when the density of small rodents was very low in spring (in years 1955 and 1958); there is an inverse correlation between losses and the density of rodents for rodent populations below about 20 rodents ha$^{-1}$. Above this value, the mortality is at a constant low value; the breeding success of the owls is not increased even when the rodent density is 70 ha$^{-1}$. We may conclude that the breeding success of tawny owls at Wytham is dependent on the density of the rodents that are their principal food in early spring, if this density is less than about 20 animals ha$^{-1}$. Above this density, losses through failure to breed are comparatively low and independent of the numbers of rodents.

*Answer 16*  $k_3$ represents the loss of eggs before hatching; it is due principally to failure of the female to incubate the eggs successfully. Figure 14 shows that the more rainy days there are during incubation, the greater are the losses of eggs. During nights with heavy rain, owls find it difficult to detect their prey because the animals move quietly over sodden leaves; several rainy nights in succession make hunting very difficult. In such conditions, the male may not be able to find enough food for himself and the incubating female; she may therefore leave the eggs to search for food for herself.

*Answer 17*  Figure 15 shows that $k_5$ is strongly density-dependent; it is almost exactly proportional to the number of owlets that are fledged. $k_5$ represents the number of young disappearing between August, when they become independent of their parents, and March, when they may start to breed themselves. Actually, some of these young may emigrate and establish themselves elsewhere; in some years, there may be an immigration of young owls from outside the estate. The greater the number of fledged owlets, the greater the number that 'disappear'.

*Answer 18*  Yes. The net result of deaths, emigration and immigration is that the population of adult tawny owls at Wytham remains almost constant.

*Answer 19*  As the number of territories increased, the average number of chicks fledged for each pair of owls decreased from about 1.2 (in 1947) to about 0.5 (1957). Values for three years (1951, 1955 and 1958) lie well below the line drawn on the graph but show a similar trend. If you check these years in Figure 11, you will find that the numbers of rodents were low in the winters. Figure 16 shows that the more adults there were present, the greater the failure in breeding performance and the fewer the number of fledged young per territory.

## 1.9 Summary

To quote Southern's own words:

1  The success of the tawny owls in Wytham in laying eggs, hatching them and in fledging young is determined by the general level of the abundance of mice and voles. This relationship finds its saturation point when the density of prey reaches 100/12 acres (approximately 20ha⁻¹, *Ed.*). Beyond this, breeding success is no further advanced.

2  The distribution of losses (i.e. failure to achieve the maximum possible) between stages of the breeding cycle shows that refusal to breed at all and failure to reach the maximum clutch size follow the total of losses closely (i.e. they represent 'key factors' in this terminology). The failure to hatch eggs, either by deserting them and/or allowing them to become chilled, also represents a heavy loss, though, if it also follows the trend of total losses, this is obscured by the sporadic, distracting influence of rainy weather. Failure to fledge the young that have been hatched, although often hailed as a notable adaptation of predators to a variable food supply, is shown to be, relatively, of little account.

3  The magnitude of these losses, depending upon the numbers of prey, is usually not sufficient to scale down the number of young produced to a point where they will replace, more or less exactly, the adults that die during a year. The adjustment takes place after fledging and during the ensuing winter, when the recently fledged young fail to secure territories within Wytham and either die there from starvation or emigrate to other areas. Nevertheless, in a few years when prey was unusually scarce, so few young were produced, that movement across the boundary of the estate was reversed.

66

# Self-Assessment Questions

## Section 20.1

### Question 1 (*Objective 1*)

Mark the following statements as either TRUE or FALSE.

(a) A community consists of many organisms all of the same species.
(b) Plants are primary producers; they 'fix' some of the energy of sunlight by the process of photosynthesis.
(c) Consumers and decomposers are heterotrophes.
(d) Carnivores are autotrophes.
(e) Energy flows through a community and then is recycled and flows through again.
(f) Most natural communities include plants, animals and micro-organisms.
(g) In any community, the number and biomass of carnivores is usually greater than the number and biomass of herbivores.
(h) Parasites are usually smaller and more numerous than their hosts.

### Question 2 (*Objective 4*)

In the following diagram, the figures are given in $10^3$ Jm$^{-2}$ per annum. Assuming that there are no other connections in the food web, fill in the values of the blanks.

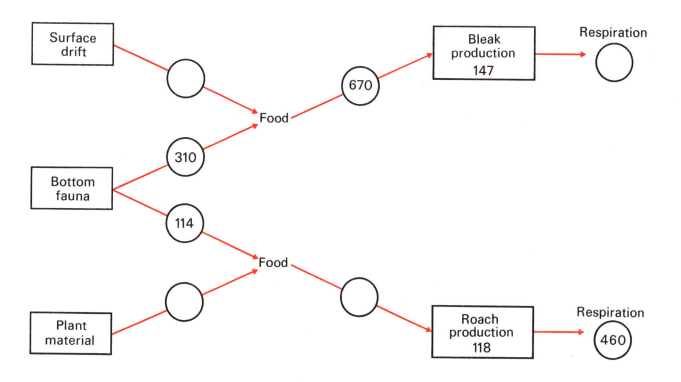

**Question 3** (*Objective 2*)

There are three series of boxes below.

(a) These boxes represent a food chain.

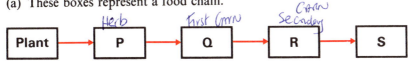

(b) These boxes represent a pyramid of numbers.

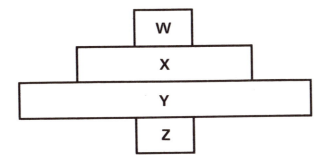

(c) These boxes represent an energy flow diagram.

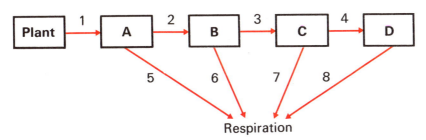

Use the information in the table below (and your general knowledge):

(a) to identify the species represented by Q, R, S and T;

(b) to identify the species represented by v, w, x, y and z;

(c) to identify the species represented by A, B, C and D and to fill in possible values for 1, 2, 3, 4, 5, 6, 7 and 8. Which, if any, of these organisms must have a further source of food energy in addition to the one in this food chain?

Table based on information for organisms in Wytham Wood, near Oxford.

| Species | Mean number per $m^2$ | Mean annual respiration $Jm^{-2}$ | Mean annual production $Jm^{-2}$ |
|---|---|---|---|
| Owl | $1.6 \times 10^{-5}$ | $21 \times 10^2$ | 12.5 |
| Predatory beetle | 60 | $12.5 \times 10^4$ | $12.5 \times 10^3$ |
| Oak-tree | $3 \times 10^{-3}$ | — | $50 \times 10^5$ |
| Shrew* | $4 \times 10^{-4}$ | $12.5 \times 10^3$ | 125 |
| Winter moth caterpillars | $10^2$ | $30 \times 10^4$ | $35 \times 10^3$ |

\* *A relative of the hedgehog; it feeds on insects.*

**Question 4** (*Objective 3*)

The matrix below contains 9 statements about metabolic capacities of groups of organisms. Choose from this matrix the statements which are true for each of the organisms in the following list; write the number of each of these statements after the organism. You may use each statement once, several times or not at all.

List of Organisms

| | |
|---|---|
| Oak-tree | Caterpillar |
| Owl | Man |
| Grass | Decomposing bacteria |

| 1   convert amino acids into mineral salts | 2   convert proteins into ammonia | 3   use the energy of sunlight |
|---|---|---|
| 4   convert nitrogen into nitrates | 5   convert carbon dioxide into sugars | 6   use the energy in sugars |
| 7   convert sugars into carbon dioxide | 8   convert mineral salts into amino acids | 9   convert sugars into amino acids |

# Section 20.2

**Question 5** (*Objective 1*)

Mark the following statements as either TRUE or FALSE.

(a) Predators are nearly always smaller than their prey.
(b) Many herbivores restrict their feeding to a small number of plant species.
(c) Detritus feeders consume only organisms smaller than themselves.
(d) The same animal occupies the same niche all over the world in all climates.
(e) The distribution of plants over the world is greatly affected by climate.
(f) Within climatic belts, the distribution of plants is not affected at all by the type of soil and other edaphic factors.

**Question 6** (*Objectives 5 and 11*)

Salmonid fishes can live in water with ice on it, but not at temperatures above 30° C.

Cyprinid fishes can live in water with ice on it and at temperatures up to 42° C.

Cichlid fishes can live only between the temperature limits of 20° to 35° C.

Given that the three statements above are true, which of the groups of fishes would you expect to find in the following waters:

(a) Lake Victoria in East Africa;
(b) Great Bear Lake in northern Canada;
(c) Lake Balaton in Hungary;
(d) fish ponds in Malaya and Nigeria;
(e) fish ponds in Sweden and Ireland;
(f) fish ponds in Israel and Poland.

**Question 7** (*Objectives 11 and 13*)

Avocado pear trees do not grow wild in the British Isles. Possible reasons for this are:

(a) that the climate is unsuitable;
(b) that the edaphic factors are unsuitable;
(c) that avocados cannot compete successfully with trees that do grow in Britain;
(d) that avocados evolved in some other part of the world and have never reached the British Isles.

Describe, using between ten and twenty short sentences, how you would investigate which of these reasons are justified explanations for the absence of avocado pears.

# Section 20.3

**Question 8** (*Objective 1*)

Mark the following statements as either TRUE or FALSE.

(a) If the numbers in a population are to remain stable, then the pre-reproductive mortality must equal the fecundity.
(b) For bacteria, which reproduce by each cell dividing into two, a constant rate of reproduction leads to exponential increase in the population.
(c) Survivorship curves always show a large decrease in numbers early in life.
(d) If population increase follows a 'logistic curve', the numbers will rise and then remain stable at the maximum value.

**Question 9** (*Objective 8*)

A population consists of 5 adult females and 5 adult males. Assuming that the generations do not overlap and that each female lays 10 eggs and all are fertilized:

How will the population of adults change if the pre-reproductive mortality is: (a) 10%; (b) 50%; (c) 90%?

(d) What pre-reproductive mortality will give a stable population?

**Question 10** (*Objective 8*)

Construct survivorship curves from the following information for three species of fish in a large river.

| Age of fish (in years) | Percentage annual mortalities | | |
| --- | --- | --- | --- |
| | Roach | Bleak | Perch |
| 0–2 | not known | not known | not known |
| 2–3 | 19 | 30 | 41 |
| 3–4 | 29 | 47 | 27 |
| 4–5 | 30 | 86 | 47 |
| 5–6 | 37 | 91 | 40 |
| 6–7 | 40 | 87 | 50 |
| 7–8 | 47 | | |
| 8–9 | 60 | | |
| 9–10 | 68 | | |
| 10–11 | 75 | | |

**Question 11** (*Objective 8*)

The following figures apply to sockeye salmon in a Canadian river system. Each female salmon lays 3 200 eggs in a gravelly shallow in the river in autumn.

640 fry (young fish, derived from these eggs) enter a lake near the shallow in the following spring.

64 smolts (older fish, survivors from the fry) leave the lake one year later and migrate to the sea.

2 adult fish (survivors of these smolts) return to the spawning grounds two and a half years later; they spawn and then die.

Calculate the percentage mortalities for sockeye salmon for each of the following periods:

(a) from the laying of the eggs in autumn to the movement of fry into the lake six months later;

(b) from entering the lake as fry to leaving the lake as smolts twelve months later;

(c) from leaving the lake as smolts to returning to the spawning grounds as adult salmon thirty months later.

Draw a survivorship curve for the sockeye salmon in this river system. What is the pre-reproductive mortality for these sockeye salmon?

## Section 20.4

**Question 12** (*Objective 1*)

Mark the following statements as either TRUE or FALSE.

(a) Herbivores may eat one plant species only.
(b) Specific parasites each attack many different host species.
(c) The numbers of herbivores are always limited by their food supply.
(d) Aggressive behaviour between animals of the same species spaces out individuals or pairs or families over the area.
(e) Key factors are always density dependent.
(f) $k$-values are the same every year.
(g) Regulating factors are always density dependent.
(h) Cycles in numbers of predators can often be explained in terms of cycles of prey numbers.

**Question 13** (*Objectives 6, 7 and 13*)

*Paramoecium* and *Didinium* are freshwater ciliate protistans; they are shown on film strip 18(b): numbers 12 and 13. *Paramoecium* feeds on bacteria, which are always present in excess in the conditions used for the following experiments; *Didinium* feeds by engulfing *Paramoecium*. Look at the three graphs shown, then answer the following questions.

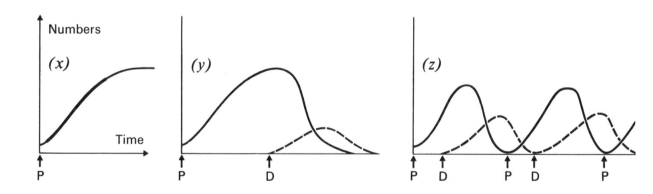

(a) Graph (*x*) shows *Paramoecium* in a culture medium without any *Didinium*. What is the term used to describe this shape of curve?
(b) How are the numbers of *Paramoecium* increasing along the part of the curve that is shown as a thicker line?
(c) Graph (*y*) shows what occurs when a few individuals of *Didinium* are added at the time shown by the arrow to a culture of *Paremoecium* similar to that shown in Graph (*x*). Describe what happens, using not more than three short sentences.
(d) Graph (*z*) shows what occurs in a culture to which small numbers of *Paramoecium* are added at times indicated by the arrows labelled P and small numbers of *Didinium* are added at times indicated by the arrows labelled D. The numbers of both species then fluctuate in a

cyclical way. Which of the following terms describes the way in which *Didinium* is acting in this experimental system:

(i) direct density dependent factor;

(ii) delayed density dependent factor;

(iii) density independent factor?

If you wish, you could attempt Exercise 4(a) on p. 57 of *Population Dynamics*.

## Section 20.5

**Question 14** (*Objective 10*)

If you were in a position to decide whether or not a certain foreign species should be introduced into this country, what types of information would you wish to have available before giving permission for its introduction? Consider this problem with special reference to (a) reindeer; (b) Canadian pondweed; (c) Colorado beetle.

Reindeer live in Scandinavia, feeding on a lichen that also grows on Scottish mountains. Canadian pondweed grows in canals and slow-flowing rivers as well as in ponds and lakes. Colorado beetle lives and feeds on potato plants; it is a native of the USA, but is widely distributed also in Europe.

## All sections

**Question 15** (*Objective 1*)

Mark the following statements as either TRUE or FALSE.

(a) Agriculture is based on mono-culture, which is contrary to the natural development of communities.

(b) Most weeds are plants that grow slowly and reproduce themselves slowly.

(c) Pests are usually carnivorous animals.

(d) Biological control typically involves using organisms that are specific predators or parasites on pests.

(e) Persistent insecticides are the best possible agents for controlling pest insects.

(f) There is plenty of protein available now for the human population.

(g) There will be no problem in producing enough food for twice the present population of the world.

(h) Problems of energy use and of waste disposal cannot be dissociated from problems of world food supply.

73

**Question 16** (*Objective 12*)

Some of the following statements about the behaviour of animals relate to this Unit, some to material in Unit 19, and some to general knowledge. Mark each statement as either TRUE or FALSE.

(a) Camouflage is a defence against predators.
(b) Camouflage allows predators to attack their prey more effectively.
(c) Camouflage is a defence against parasites.
(d) Song is part of courtship behaviour of birds only.
(e) Song is part of territorial behaviour of birds only.
(f) Courtship behaviour is one mechanism isolating members of one species from those of closely related species.
(g) Related species of birds usually look very much alike and have very similar courtship displays.
(h) Related species of birds nearly always have different courtship displays.
(i) Many herbivorous animals live in herds or large groups.
(j) Many carnivorous animals are solitary or live in small groups.

**Question 17** (*Objectives 9 and 13*)

In 1965, the world population of grey seals was estimated as 46 000; of these, 36 000 live round the British Isles. About one-tenth of the British grey seals are based on the Farne Islands (off the coast of Northumberland) for which the following figures are available:

(a) the number of calves produced annually was about 100 in the 1930s; about 600 in 1952; and more than 1 000 in 1962;
(b) the seals 'calve' on four islands only—here is information about two of these islands.

|  | *Staple* | *Brownsman* |
|---|---|---|
| Number of calves per 100 yds of accessible beach | 77 | 14 |
| Percentage of calves dying during: | | |
| first half of breeding season | 16.4 | 8.7 |
| second half of breeding season | 26.4 | 11.8 |

Seals are unselective carnivores, their diet includes the following species of fish: salmon, cod, plaice, sand eels, lumpsuckers.
Local fishermen believe that seals prefer lumpsuckers to salmon.
Lumpsuckers eat lobsters.

Use the information given above to construct three arguments *in favour* of action to reduce the number of seals breeding on the Farne Islands *and* three arguments *against* such action.

**Question 18** (*Objectives 5, 9 and 13*)

The lowest levels of dissolved oxygen as cm³ per litre which can be tolerated by the following organisms are:

*Tubifex* 0.2; *Erpobdella* 2.0; *Chironomus riparius* 0.5; *Limnaea pereger* 1.7; *Asellus* 1.8; *Baetis rhodani* 3.0; *Gammarus* 3.2; *Ancylus* 2.8; *Phoxinus* 6.0; *Heptagenia* 5.8.

Direction of flow ➝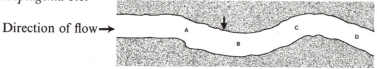

In the river above, the fauna at sites A, B, C and D were as follows:

| A | B | C | D |
|---|---|---|---|
| *Phoxinus, Heptagenia, Gammarus, Ancylus, Baetis, Limnaea* | *Tubifex* only | *Chironomus* only | *Chironomus, Erpobdella, Asellus, Limnaea* |

Suggest what happened to the river at the point marked by the arrow. What does the change in fauna between B and D indicate?

**Question 19** (*Objectives 9 and 13*)

Charles Darwin quoted the following figures:

100 heads of red clover produced 2 700 seeds, but the same number of heads protected from the attention of humble-bees produced not a single seed. He quoted Colonel Newman as saying 'more than two-thirds of them (i.e. humble-bees) are destroyed all over England' by field-mice, and also 'near villages and small towns I have found the nests of humble-bees more numerous than elsewhere'.

Darwin put forward the following hypothesis:

'. . . the presence of a feline animal in large numbers in a district might determine, through the intervention first of mice and then of bees, the frequency of certain flowers in that district'.

Read the following statements, then arrange some of them in a sequence to support Darwin's hypothesis.

Try to arrange some of the statements in a sequence to form a different hypothesis.

(a) Elderly ladies who keep cats choose to live in villages rather than in solitary cottages.

(b) Farmers often have granaries infested with rats and mice.

(c) Cats kill field-mice.

(d) Farmers cut down trees except sometimes along field boundaries.

(e) Humble-bees nest in burrows in short grass.

(f) Owls feed on small rodents including field-mice.

(g) Many houses in villages have gardens with lawns and trees.

(h) Owls wait in trees and pounce when they hear a rustle of small animals in the grass.

(i) Farmers keep cats to control rats and mice in their buildings.

# Self-Assessment Answers and Comments

## Question 1

(a) False, see 20.0.
(b) True, see 20.1.
(c) True, see 20.1.
(d) False, see 20.1.
(e) False, see 20.1.1 and *Ecological Energetics* 1.6 and 2.4.
(f) True, see 20.1.2.
(g) You should have found it impossible to answer this question simply as true or false. The *biomass* of carnivores is generally less than that of herbivores (see *Ecological Energetics* 2.4); but in some communities the herbivore biomass is small and the carnivores eat detritus feeders (decomposers) as described in *Ecological Energetics* 3.2 and 3.3. The number of carnivores is generally smaller than the number of herbivores, but this is not true where the bulk of carnivore food is provided by decomposers.
(h) True, see *Ecological Energetics* 2.4.

## Question 2

The roach food intake must supply the energy for respiration and production: $460 + 118 = 578 \times 10^3$ J m$^{-2}$ per annum.

This roach food includes bottom fauna and plant material. The amount of plant material must be: total food $-$ bottom fauna $= 578 - 114 = 464 \times 10^3$ J m$^{-2}$ per annum.

Bleak food includes bottom fauna and surface drift. The amount of surface drift must be: total food $-$ bottom fauna $= 670 - 310 = 360 \times 10^3$ J m$^{-2}$ per annum. This bleak food intake supplies energy for production and respiration so the amount of energy dissipated in respiration is: total food $-$ production $= 670 - 147 = 523 \times 10^3$ J m$^{-2}$ per annum.

If you feel uncertain about the basis for these calculations, read *Ecological Energetics* 3.2 and the commentary on it in the text under 20.1.1.

## Question 3

This is an exercise based on understanding ecological pyramids and energy flow diagrams (see *Ecological Energetics* 2.4 and 3.2 and the commentary in this text under 20.1.1).

(a) P (herbivores) = winter moth caterpillars; Q (first carnivores) = predatory beetles; R (second carnivores) = shrews; S (top carnivore) = owl.
(b) z (plants) = oak-trees; y (herbivores) = winter moth caterpillars; x (first carnivores) = predatory beetles; w (second carnivores) = shrews. If w is taken to include all carnivores above the first level, then owls are also included.
(c) *A* (herbivores) = winter moth caterpillars.
Arrow 5 represents their respiration: $30 \times 10^4$ J m$^{-2}$.
Arrow 1 represents their consumption of energy, which equals their production + their respiration: $35 \times 10^3 + 30 \times 10^4 = 33.5 \times 10^4$ J m$^{-2}$.

76

$B$ (first carnivores) = predatory beetles.

Arrow 6 represents their respiration: $12.5 \times 10^4$ J m$^{-2}$.

Arrow 2 represents their consumption of energy, which equals their production + their respiration:
$$12.5 \times 10^4 + 12.5 \times 10^3 = 13.75 \times 10^4 \text{ J m}^{-2}.$$

$C$ (second carnivores) = shrews.

Arrow 7 represents their respiration: $12.5 \times 10^3$ J m$^{-2}$.

Arrow 3 represents their consumption of energy, which equals their production + their respiration: $125 + 12.5 \times 10^3 = 12.63 \times 10^3$ J m$^{-2}$.

$D$ (top carnivores) = owls.

Arrow 8 represents their respiration: $21 \times 10^2$ J m$^{-2}$.

Arrow 4 represents their consumption of energy, which equals their production + their respiration: $12.5 + 21 \times 10^2 = 21.13 \times 10^2$ J m$^{-2}$.

Compare the production of each organism in the food chain with the consumption of the organism which feeds on it—if an organism is the sole source of food for the next consumer, then the production of the food organism will be at least one order of magnitude greater than the consumption of the organism that eats it. If this is not so, the population of the food organism will probably decrease in biomass and become extinct.

Oak-tree production is $50 \times 10^5$, winter moth caterpillar consumption is $33.5 \times 10^4$, so these caterpillars could feed entirely on oak-trees.

Winter moth caterpillar production is $35 \times 10^3$; beetle consumption is $13.8 \times 10^4$ so the winter moth population could not support the beetles. The predatory beetles must have other sources of food energy—in fact, they eat a variety of species of insects.

Beetle production is $12.5 \times 10^3$; shrew consumption is $12.6 \times 10^3$; so the shrews must have other sources of food in addition to predatory beetles.

Shrew production is only 125, but owl consumption is $21 \times 10^2$; so the owls must have sources of food in addition to shrews. (In fact, owls usually do not eat shrews; they prey on other small mammals such as field-mice, voles and moles—see Appendix 1.)

**Question 4**

If you cannot answer this question, you should read 20.1 and 20.1.2 again.

From the metabolic point of view, the organisms fall into three groups:
(a) photosynthetic autotrophes: oak-tree and grass. Statements 3, 5, 6, 7, 8 and 9 are true for these.
(b) heterotrophes that consume living or freshly killed organisms: caterpillars (herbivores), owl (carnivore) and man (omnivore). Statements 2, 6 and 7 are true of these. You may have put man in a special category, since some of these conversions can be carried out in laboratories or factories as a result of human activity (e.g. 4) but this is not part of 'human metabolism'.
(c) heterotrophes that are decomposers: decomposing bacteria. Statements 1, 2, 6 and 7 are true of these. Statement 4 is true of a different group of bacteria, the 'nitrogen-fixing bacteria'.

## Question 5

(a) False, see 20.2.
(b) True, see 20.2.1.
(c) False, see 20.2.
(d) False, see 20.2.1.
(e) True, see 20.2.2.
(f) False, see 20.2.2.

## Question 6

(a) The temperature in Lake Victoria is usually close to 25° C. This is well within the tolerance of cyprinid and cichlid fishes (and both groups are present there), but it is dangerously close to the upper limit for salmonid fishes (which are not native there).

(b) Great Bear Lake freezes in the winter and remains cool through the summer. Cichlid fishes could not live there, but salmonid and cyprinid fishes could.

(c) Lake Balaton freezes in the winter and becomes very warm in the summer. Cichlid fishes could not survive the winter and salmonids could not survive hot summers; it should support (as it does) cyprinid fishes.

(d) Fish ponds in Malaya and Nigeria contain warm water all through the year. This is ideal for cichlid and cyprinid fishes, but salmonids could not survive.

(e) The water temperature may fall to freezing in the winter in Sweden and Ireland, but it remains cool in the summer. These ponds are ideal for salmonid fishes (such as trout) and cyprinids can also live in them; cichlids would not survive through most of the year.

(f) In Israel and Poland, the water becomes very cold and may freeze in winter; it becomes very warm in summer. Cichlids would not survive through the winter nor salmonids through the summer; these are ponds for rearing cyprinid fishes (such as carp).

## Question 7

To investigate whether the climate (a) and edaphic factors (b) in Britain are suitable for avocado trees or not, you could try to cultivate avocado nuts in different parts of Britain. You would use local soils under greenhouse conditions (with temperature and light adjusted appropriately—see under (d)) to investigate edaphic factors. You would grow the plants out-of-doors but in prepared compost or humus to investigate climate. You would grow the plants out-of-doors and in the local soil to investigate any interaction between soil and climate.

You would not be able to investigate (c) unless the plants could survive both the British climate and in at least one type of British soil. If they could survive, then you could plant them out with seedlings of trees and compare survival under these conditions with survival under non-competitive conditions.

To investigate (d) you would find out from books and scientific papers where avocados grow wild and whether there is information to show that they are native to those areas or have been introduced. Eventually you might be able to locate a limited part of the world that represents their natural distribution. You would know as a result of your researches the type of climate in which avocados grow or can be cultivated and you could use this information in your greenhouse experiments.

Read 20.2.2 if you need to remind yourself about the factors affecting the distribution of plants.

### Question 8

(a) False, see 20.3. Births minus deaths must equal parent numbers.
(b) True, for the early growth of bacteria in a culture medium; the growth of the population later declines giving a 'logistic' curve, see 20.3.1.
(c) False, since some curves, such as that for man, do not follow this pattern; other curves do show this early decrease in numbers, see 20.3.
(d) True, see 20.3.1.

### Question 9

The 5 females will lay 50 fertilized eggs.

(a) if the pre-reproductive mortality is 10 per cent, the offspring generation will be 45
(b) if the pre-reproductive mortality is 50 per cent, the offspring generation will be 25
(c) if the pre-reproductive mortality is 90 per cent, the offspring generation will be 5.

So for (a), the offspring population will be 4.5 times the parent population.

So for (b), the offspring population will be 2.5 times the parent population.

So for (c), the offspring population will be 0.5 times the parent population.

(d) For a stable population of 10 adults in each generation, 10 offspring must survive from the 50 eggs. Since 40 offspring must die, the mortality must be 80 per cent.

**Question 10**

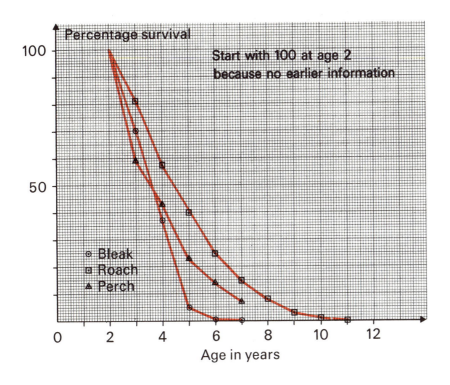

**Question 11**

(a) Out of 3 200 eggs, 640 survive, so 2 560 die—a mortality of 80%.

(b) Out of 640 fry, 64 survive, so 576 die—a mortality of 90%.

(c) Out of 64 smolts, 2 survive, so 62 die—mortality of about 97%.

The total pre-reproductive mortality for salmon is 3 198 out of 3 200—99.97%.

3 200 eggs→640 fry→64 smolts→2 adult salmon.

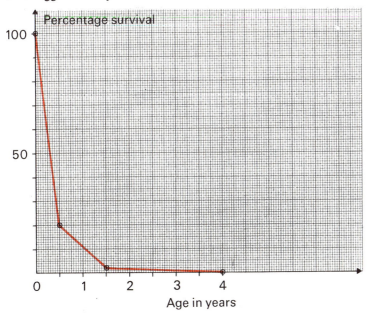

**Question 12**

(a) True, see 20.2 and 20.4.3.
(b) False, the meaning of 'specific' is that this type of parasite is limited to one host species or a small number of related species, see 20.4.2.
(c) False, see 20.4.3.
(d) True, see 20.4.
(e) False, usually they are independent of density, see 20.4.1.
(f) False, see 20.4.1.
(g) True, if delayed density dependent factors are included, see 20.4.2.
(h) True, see 20.4.4.

**Question 13**

(a) This is a 'logistic' curve—compare it with Figure 3.3 of *Population Dynamics*.
(b) Exponentially, see 20.3.1.
(c) *Didinium* consumes *Paramoecium* faster than the latter can reproduce themselves. *Didinium* increases in numbers until it is limited by its decreasing food supply; then the numbers begin to fall. Both *Paramoecium* and *Didinium* become extinct.
(d) It is acting as (ii) a delayed density dependent factor. The peaks and troughs of *Didinium* numbers follow the peaks and troughs of *Paramoecium* numbers regularly.

**Question 14**

You would wish to know the possible effects of introducing the foreign species into British communities of plants and animals. The foreign species could affect native species (cultivated and wild) through food web relations (by consumption or competing for a food supply or altering the food web in other ways) or through the introduction with it of pests or diseases that might have a devastating effect on native organisms. The latter problem could be mitigated by quarantining the organism—cultivating it under isolation so as to clear it of pests and diseases as far as possible. The problem of effects through food web relations could be clarified by considering the species' food web relations in its native environment. To take our three examples.

(a) Reindeer are herbivores with a limited food preference—they would not compete with native sheep or cattle or native deer. Their activity might lead to soil erosion through eating out of lichens—the probability of this could be judged by comparing the environment in Scandinavia with that in Scotland. It is unlikely that native predators would be able to tackle such large animals. Reindeer are now herded near Aviemore in Scotland.

(b) Canadian pondweed was introduced into Britain in 1842; it propagated itself so vigorously that it soon choked drainage ditches in the Fens and also some rivers and canals. After flourishing at a nuisance level between 1850 and 1880, it began to propagate less luxuriantly (the cause for this is not understood). Now it is a very common plant and

is considered useful since it adds much oxygen to the water and supports many animals that are valuable foods for fishes. An experimental introduction in conditions where the plant could not spread to other waters might have revealed its undesirable (in the short term) vigour but not its long-term usefulness.

(c) The colourful Colorado beetle is a pest in the USA and in Europe. A very brief investigation of the devastation it causes to potato fields in both continents would be sufficient to convince any reasonable person that it should be kept out of the British Isles. In fact, there are regulations forbidding its introduction here. Small numbers of live beetles have occasionally been found in southern counties in summer; immediate efforts to eradicate them before the species could become established have so far been successful.

## Question 15

(a) True, see 20.6.
(b) False, see 20.6.
(c) False, most pests are probably herbivores, see 20.6.1.
(d) True, see 20.6.1.
(e) False, see 20.6.1.
(f) False, see 20.6.2.
(g) False, there are already problems with the present population, see 20.6.2.
(h) True, see 20.6.2.

## Question 16

(a) True—provided that the predators hunt mainly by sight; remember the peppered moth shown in TV 19.
(b) True—provided that the prey detect predators by sight; think of lions concealed on dry grassland and tigers in forests.
(c) True—only if parasites find their hosts by sight; this is unusual—most parasites are probably attracted by scents.
(d) False—it is part of courtship but also usually part of other aspects of behaviour.
(e) False—it is part of territorial behaviour but also usually part of courtship behaviour also.
(f) True—see 19.7.2.
(g) False—they may resemble each other but their courtship behaviour is usually markedly different—see 19.7.2. (Consider ducks in parks).
(h) True—this is the converse of (g).
(i) True—think of antelopes, zebra, caribou and buffaloes.
(j) True—think of lions and eagles.

**Question 17**

Possible arguments in favour of reducing the number of **seals** that breed on the Farne Islands are as follows.

1  Seals eat salmon, cod and plaice, all valuable fishes (they also damage salmon nets by taking fish out of them)—and thus reduce our potential food supply.

2  The number of seals has increased so markedly since the 1930s that it appears that any natural balance has been upset and the numbers may just go on increasing indefinitely.

3  Where there are many seals, there is a higher mortality of calves; to allow the total number of seals to go on increasing will lead to greater suffering among calves; to reduce the number of seals will reduce suffering.

Possible arguments against reducing the number of seals that breed on the Farne Islands are as follows.

(a) The grey seal is a rare species; 75 per cent of the world population lives in British waters; we have a responsibility to protect them so it would be wrong to kill any.

(b) Reducing the number of seals will mean more lumpsuckers and hence fewer lobsters; lobsters are a prized delicacy which should be protected.

(c) Since the proportion of calves dying is higher where the seals are more crowded, reducing the number of breeding seals may not in fact reduce the number of adults because the total number of calves surviving may be the same as at present. This death of calves may be the regulating factor—interference with this may lead to unexpected results.

**Question 18**

The fauna at point A includes animals which require high levels of dissolved oxygen (e.g. *Phoxinus* 6.0, *Heptagenia* 6.8), so here there must be at least 7 cm³ of dissolved oxygen per litre.

At B, only *Tubifex* is present, suggesting that the level of dissolved oxygen must be between 0.2 and 0.5 cm³ per litre. Therefore something happened to the river at the point marked by the arrow resulting in a fall in dissolved oxygen. At D, the level must be at least 2.0 cm³ per litre (since *Erpobdella* lives there). Therefore as it flows from B to D, the amount of dissolved oxygen in the water increases from 0.5 to 2 cm³ per litre.

These are the kind of changes in dissolved oxygen and in river fauna that occur when poorly treated domestic sewage enters a river (at the point marked by the arrow) and causes pollution from which the river gradually recovers as it flows on.

## Question 19

Darwin assumes that the number of field-mice determines the number of bees and that clover plants will grow only where bees have visited the flowers. If there are more bees near villages and small towns, this implies that there are fewer field-mice there. Darwin suggests that the low numbers of field-mice are the result of the presence of many cats near villages and small towns.

Of the nine statements, (e) followed by (g) account for the presence of humble-bees near villages; (c) followed by (a) account for the absence of field-mice near villages; (i) implies that there could be many cats in the countryside but (b) explains that these cats probably hunt in granaries and do not molest the field-mice that eat the humble-bees. So Darwin's hypothesis is supported by the observations in statements (e), (g), (c), (a), (i) and (b).

An alternative hypothesis is supported by statements (e), (g), (f), (h), (d) and (g) again. It is that the number of field-mice is controlled by owls; these birds hunt from trees; there are few suitable trees in the countryside because farmers cut them down, but there are suitable trees in village gardens; hence there are fewer field-mice near villages and therefore (from Colonel Newman's statement) more humble-bees and so probably more clover (from Darwin's own statement).

# Legends for Film Strips

## Film strip 19/20(a)

1 Eggs and larva of the fruitfly *Drosophila melanogaster* ($\times 60$). One egg is shown seen from the side (with two hornlike spiracles (breathing tubes) sticking up); the other egg is seen from above with the two spiracles flattened outwards. The larva is a small maggot.

2 Two puparia of *Drosophila melanogaster* ($\times 20$). One seen from the side and the other from above. The adult insect is developing inside the hard case of the puparium.

3 Anaesthetized adult fruitflies, *Drosophila melanogaster* ($\times 10$). Note the general resemblance to small houseflies and the characteristic colour pattern. Compare these flies with the drawings in your Home Experiment notes for Unit 20.

4 Adults of two common species of fruitfly ($\times 10$). On the right is *Drosophila melanogaster* (compare with photograph 3) and on the left is *Drosophila subobscura*. Note the differences in size and colour pattern between these two species. Compare the photograph with the drawings in your Home Experiment notes for Unit 20.

5 Two parasites of the fruitfly *Drosophila*. These were both bred from wild *Drosophila* populations reared as described in your Home Experiment notes for Unit 20. Compare these two specimens with the photographs of the adult fruitflies (3 and 4): note the differences in body shape (the parasites are both wasps with slender waists), in wing shape and pattern and in length of antennae (feelers). A live specimen of *Phaenocarpa* (5b) is shown in the TV programme for Unit 20; the other parasite (5a) is called *Pseudeucoila*. If you collect either of these two insects while carrying out your Home Experiment for Unit 20, preserve the specimen and inform us.

6 Mouse chromosomes (*Unit 19*) ($\times 1\,000$) shortly before the first meiotic division. Note that each is divided into two chromatids. Attempt to count the number of chromosomes.

7 Mouse chromosomes ($\times 1\,000$). The chromosomes have become associated in homologous pairs, each pair comprising four chromatids. Count the number of pairs.

8 A further enlargement of one homologous pair of chromosomes, clearly showing chiasma formation and the physical crossing-over of the chromatids.

## Film strip 20(b)   Photographs of British vegetation

1 Oakwood on clay soil, photographed in May 1970. Note the bluebells and other herbs under the trees, which are not yet in full leaf.

2 Oakwood on dry sandy soil, photographed in June 1970. Note the dense undergrowth of bracken.

3 Open pinewood. Note the open glades and dense growth of bracken.

4 Dense pinewood. Note the close spacing of the trees and the absence of undergrowth.

5 Lullington Heath, Sussex: photograph taken on 23 March 1954. Note the rabbit burrows surrounded by bare chalk, the short turf and the elder bushes. Rabbits were common here.

6 Lullington Heath, Sussex: photograph taken on 21 February 1967. This is the same area as that shown in (5). Note the dense sward of grass and the tangle of bramble bushes in front of the elder bush. Rabbits had died of myxomatosis.

7 Old Winchester Hill, Hampshire: photograph taken on 10 August 1954. Note the presence of chalk lumps forming scree and the large numbers of ragwort plants (with yellow flowers). Rabbits were common here.

**8** Old Winchester Hill, Hampshire: photograph taken on 13 August 1956. This is the same view as that in (7). Note that the turf has grown over most of the scree; there are no ragwort plants. Rabbits had died of myxomatosis.

### Film strip 20(c)   Moths, caterpillars and their parasites

**1** A 'wingless' female moth on the trunk of an oak-tree. Note the short wings, long legs and abdomen swollen with eggs. This is the Scarce Umber *Erannis aurantiaria*; it closely resembles the winter moth (2).

**2** Winter moth, *Operophtera brumata*: wingless female and winged male photographed by flash at night in November on the trunk of an oak-tree.

**3** Winter moth caterpillar photographed in May on an oak twig.

**4** *Cyzenis*, a fly whose larva is a parasite of the winter moth caterpillar. On the left, a dead 'Museum' specimen, on the right, a live fly. Note the similarity to houseflies. (Ignore the word Phygadeuon.)

**5** *Cratichneumon*, a wasp whose larva is a parasite of the winter moth pupa. Male (left) and female (right), each mounted above the winter moth pupal case (chrysalis) from which it emerged.

**6** November Moth, *Oporinia dilutata* caterpillar on an oak twig with the young leaves beginning to unfold in April. The adult moth is about in October and November.

**7** Scarce Umber caterpillar photographed in May on an oak leaf.

**8** Scarce Umber, *Erannis aurantiaria*: wingless female and winged male on the trunk of an oak-tree in November. Note the similarity to the winter moth (2).

## Acknowledgements

Grateful acknowledgement is made to the following sources for material used in this Unit:

TEXT

THE ZOOLOGICAL SOCIETY OF LONDON for H. N. Southern in *Journal of Zoology*, Vol. 162, Part 11, October, 1970.

ILLUSTRATIONS

PROFESSOR G. C. VARLEY for Figs. 7, 8, 9; BLACKWELL SCIENTIFIC PUBLICA-TIONS for Fig. 10 from G. C. Varley and G. R. Gradwell in T. R. E. Southwood (ed.), *Insect Abundance*; THE ZOOLOGICAL SOCIETY OF LONDON for Figs. 11, 12, 13, 14, 15, 16 from *Journal of Zoology*, Vol. 162, Part 11, October, 1970.

**Notes**

## S.100—SCIENCE FOUNDATION COURSE UNITS

The Open University

Science Foundation Course Unit 19

## EVOLUTION BY NATURAL SELECTION

*Prepared by the Science Foundation Course Team*